TRANSLATION AND LOCALIZATION

Detailed yet accessible, *Translation and Localization* brings together the research and insights of veteran practicing translators to offer comprehensive guidance for technical communicators. The volume begins with the fundamentals of translation before leading readers through the process of preparing technical documents for translation. It then presents the broader area of localization, again beginning with its key competencies. Concluding chapters examine the state of the field as computers take on more translation and localization work. Featuring real-life scenarios and a broad range of experienced voices, this is an invaluable resource for technical and professional communicators looking to expand into international markets. This book will be of interest to students of ethnic conflict, Asian politics, and security studies.

Bruce Maylath is Professor of English at North Dakota State University (USA) and Director of the university's program in Upper-Division Writing.

Kirk St.Amant is Professor and Eunice C. Williamson Endowed Chair in Technical Communication at Louisiana Tech University (USA), Adjunct Professor of International Health and Medical Communication with the University of Limerick (Ireland), and Guest Professor of Usability and Design with Southeast University (China).

ATTW Book Series in Technical and Professional Communication

Tharon Howard, Series Editor

Technical Communication After the Social Justice Turn: Building Coalitions for Action
Rebecca Walton, Kristen R. Moore and Natasha N. Jones

Translation and Localization: A Guide for Technical and Professional Communicators
Bruce Maylath and Kirk St.Amant

Editing in the Modern Classroom
Michael J. Albers and Suzan Flanagan

Creating Intelligent Content with Lightweight DITA
Carlos Evia

Involving the Audience: A Rhetorical Perspective on Using Social Media to Improve Websites
Lee-Ann Kastman Breuch

Rhetorical Work in Emergency Medical Services: Communicating in the Unpredictable Workplace
Elizabeth L. Angeli

Citizenship and Advocacy in Technical Communication
Godwin Y. Agboka and Natalia Mateeva

Communicating Project Management
Benjamin Lauren

For additional information on this series please visit www.routledge.com/ATTW-Series-in-Technical-and-Professional-Communication/book-series/ATTW, and for information on other Routledge titles visit www.routledge.com.

TRANSLATION AND LOCALIZATION

A Guide for Technical and Professional Communicators

Edited by Bruce Maylath and Kirk St.Amant

Routledge
Taylor & Francis Group
NEW YORK AND LONDON

First edition published 2019
by Routledge
52 Vanderbilt Avenue, New York, NY 10017

and by Routledge
2 Park Square, Milton Park, Abingdon, Oxon, OX14 4RN

Routledge is an imprint of the Taylor & Francis Group, an informa business

Library of Congress Cataloging-in-Publication Data
Names: Maylath, Bruce, editor. | St.Amant, Kirk, 1970– editor.
Title: Translation and localization : a guide for professional and technical communicators / Bruce Maylath, Kirk St.Amant [editors].
Description: London ; New York, NY : Routledge, 2019. |
Series: ATTW book series in technical and professional communication
Identifiers: LCCN 2019000515| ISBN 9781138319936 (hardback) | ISBN 9781138319943 (pbk.)
Subjects: LCSH: Technology—Translating. | Translating and interpreting. | Communication of technical information.
Classification: LCC T11.5 .T58 2019 | DDC 601/.4—dc23
LC record available at https://lccn.loc.gov/2019000515

ISBN: 978-1-138-31993-6 (hbk)
ISBN: 978-1-138-31994-3 (pbk)
ISBN: 978-0-429-45367-0 (ebk)

Typeset in Minion
by codeMantra

To the memory of Dan Riordan: In his four decades guiding students, Dan exemplified the highest quality in teaching technical communication and fostering its development around the world.

To the memory of Victoria Mikelonis, for her career-long advocacy of the importance of language, culture, and translation in teaching technical communication, both in the United States and in other nations.

CONTENTS

CONTRIBUTORS

Co-Editors

Bruce Maylath is a professor of English at North Dakota State University (USA), where he directs the program in Upper-Division Writing, teaches courses in linguistics, and researches translation issues in international technical communication. He is the co-founder and the coordinator of the Trans-Atlantic and Pacific Project, which links writing, usability testing, and translation classes at universities across four continents.

Kirk St.Amant is a professor and Eunice C. Williamson Endowed Chair in technical communication at Louisiana Tech University (USA), where he is also a research faculty member with the University's Center for Biomedical Engineering and Rehabilitation Science (CBERS). St.Amant is also an adjunct professor of international health and medical communication with the University of Limerick (Ireland) and a guest professor of usability and design with Southeast University (China). He researches how cognition affects usability and the design of technology with a focus on international health and medical contexts and international online education.

Authors

Patricia Minacori is an associate professor at the University of Paris—Diderot (France). A former technical translator, she now serves as director of a department in charge of translation and technical communication and as head of two curricula in technical communication. She has taught technical and scientific translation for almost 30 years in graduate degree programs. Her research focuses on translation project management, complexity according to

Edgar Morin, exchanges between technical communicators and translators, and translation assessment.

Birthe Mousten holds a Ph.D. in translation/localization, M.A. in English, B.A. in German, diploma in marketing, and is a certified translator/interpreter of English/Danish. Since 1992, she has lectured at the four major Danish universities: Copenhagen University (KU), Aalborg University (AAU), University of Southern Denmark (SDU), and now primarily Aarhus University (AU). Her specialty areas include 1) translation: LSP, legal and technical translation, LGP, literary translation; 2) communication: intercultural communication, negotiation; 3) talks: seminars, conferences and workshops; 4) research: translation/localization, didactics in translation, lexicography, and ad-hoc corpus linguistics.

Maria Teresa Musacchio is a professor of English language and translation in the Department of Linguistic and Literary Studies at the University of Padua (Italy), where she coordinates the postgraduate course in modern languages for international communication and cooperation. Her research focuses, in a functional, pragmatic and cultural perspective, on descriptive and applied translation studies issues in specialized translation and the related fields of special languages, terminology, and translation pedagogy. Since 2002, her corpus-based and corpus-driven investigations have specifically explored communication, discourse, and register analysis in the language and translation of science popularization. From 2014 to 2017, she was head of the Padua unit of the EU FP7 Project Slándáil—Security System for Language and Image Analysis—focusing on the use of language and translation to empower emergency response through the use of social media.

Sharon O'Brien is an associate professor at the School of Applied Language and Intercultural Studies, Dublin City University (Ireland). Her research interests focus on the topics of translation technology, controlled authoring, cognition, ergonomics, and translation in disaster prevention and management. She teaches in B.A. and M.A./M.Sc. programs related to translation, focusing mainly on translation technology, localization, and research methods.

Giuseppe Palumbo is an associate professor of English language and translation at the University of Trieste (Italy), where he is also head of the bachelor degree program in language mediation for the legal professions. He holds a Ph.D. from the University of Surrey (UK) and specializes in technical and scientific translation and the use of technology in professional translation. He has published on corpus-based analyses of English, terminology, the design of translator training curricula, and the creation of online translation teaching materials. He is the author of *Key Terms in Translation Studies* (Continuum, 2009).

Dan Riordan, professor emeritus at the University of Wisconsin—Stout (USA), taught English there from 1970 to 2010. During that time, he designed and helped launch Wisconsin's first program in technical communication, as well as helping found and then directing the Nakatani Teaching and Learning Center. He was the author of *Technical Report Writing Today* in its 2nd through 10th editions. A past president of the Council for Programs in Technical and Scientific Communication, he was honored in 2003 by the Society of Technical Communication with its Jay R. Gould Award for Excellence in Teaching.

Federica Scarpa is professor of English language and translation at the Department of Legal, Language, Interpreting, and Translation Studies of the University of Trieste (Italy), where she has taught specialized translation between English and Italian since 1991. She was the coordinator of a Ph.D. program in interpreting and translation studies (2009–2015) and director of a master's degree in legal translation (2012–2016), both at the University of Trieste. The second edition of her book on specialized translation (Hoepli, Milan, 2008) was translated into French and published by Ottawa University Press in 2010.

Tomáš Svoboda is head of the German department at the Institute of Translation Studies, Charles University Prague (Czech Republic). He holds a Ph.D. in translation studies from Charles University, and his publications include topics such as translation technology, the future of the translation profession, institutional translation, translation quality, and translation history. He is a board member of the EMT (European Master's in Translation) university network and executive board member of the Czech Union of Translators and Interpreters.

Paola Valli holds a Ph.D. from the University of Trieste (Italy) in translation and interpreting studies, which focused on translation technologies, translation problems, and translation process research. She advanced her education with study visits at the University of Iowa, Copenhagen Business School, and Saarland University and gained further experience in the translation industry working at Lionbridge/CLS 4-Text in Berlin. She also worked as a translator in Rome and at the European Parliament in Luxembourg. In a product manager capacity at TAUS, she later helped develop the TAUS Quality Dashboard, an industry-shared platform that visualizes translation quality and productivity data.

Pavel Zemliansky, Ph.D. is an associate professor in the Academic Writing Program at Oslo Metropolitan University in Oslo, Norway, where he teaches undergraduate and graduate courses in academic writing and rhetoric. His research has appeared in the journals *IEEE Transactions on Professional Communication*, *The Journal of Technical Writing and Communication*, numerous edited collections, and other outlets. He is also editor or co-editor of several academic collections, including the influential open access composition resource *Writing Spaces*.

ACKNOWLEDGMENTS

This book would not have been possible without the knowledge and patience of its talented contributors; the perceptive and helpful recommendations of blind reviewers; the encouragement of Tharon Howard, editor of the ATTW Series in Technical Communication; the guidance of Nicole Salazar, editor at Routledge/Taylor & Francis; and the support of our spouses and families. We thank them all profusely.

SERIES EDITOR FOREWORD

It has been my pleasure to include Bruce Maylath and Kirk St.Amant's edited collection *Translation and Localization: A Guide for Professional and Technical Communicators* in the ATTW Series in Technical and Professional Communication. Sadly, our field (particularly in the United States) has suffered from what Bruce calls "a cult of monolingualism." There are certainly organizations like NAFSA which provide useful and important work on translation and even machine translation in a generalized way, but there is surprising lack on current research and scholarship that speaks directly to the needs of technical and professional communicators. However, as one reviewer of the manuscript put it: "Together, the chapters in this collection provide an exciting up-to-date overview of localization and translation as complementary activities that hinge on the constant interactions between humans and machines. Individually, each chapter fulfills the purpose it describes, while pointing to the contents of complementary chapters."

Today, our Technical and Professional Communication classrooms are uniquely positioned to prepare our students to become leaders in the translation and localization activities of their companies. The chapters in this collection are all intended to move our courses beyond merely training students to "translate" scientific and technical content into "a language" which English-speaking laypersons can understand, and instead, the chapter authors help us see ways to make our courses pathways to citizenship in both local and global contexts. The chapters in this collection all challenge us to go beyond producing graduates who develop the texts that explain how a given process works or how a particular technology operates and encourage both practitioners and faculty to situate the usability and accessibility of information products into the realm of communication practices at the local level.

Maylath and St.Amant have assembled some of the best and brightest scholars in the field as contributors to this volume. As a result, this book provides an important contribution to our field's understanding of the theoretical frameworks and empirical research methodologies which are being used to approach issues such as web localization, technical translation, "controlled" language, cultural competencies, virtual teams, and machine translation. In our "global economy" and in a world where Internet access connects billions of people across international borders, it's easy to understand why TPC practitioners and academics need the introduction to emerging trends in translation and localization this book offers.

Again, I'm thrilled that this collection is part of the ATTW Series in Technical and Professional Communication. The topic is totally consistent with and wholly appropriate for the ATTW Series and provides exactly the blend of solid scholarship and pedagogical application that we're seeking in the Series.

Dr. Tharon W. Howard
Editor, ATTW Book Series in Technical and
Professional Communication
February 1, 2019

FOREWORD

Pavel Zemliansky

I'd like to begin with an anecdote. A long time ago and in a land far, far away, I used to work as a translator and interpreter. I was often hired both to translate business and technical documents and to provide oral interpretation at business meetings, conferences, and other professional events. One day, I was working as an interpreter at a three-day business meeting between executives of a large elevator-manufacturing company and a client who wanted to buy a lot of elevators from them. On day three of the meeting, during a fancy dinner at an expensive restaurant, I was asked to interpret into English an impromptu pitch from the client, who was hoping to close the deal right then and there. The client, a gray-haired man in his sixties, began a sentence, got through it halfway, then looked at me. I looked back at him, waiting for him to finish the sentence. He continued to look at me then said, "Please translate what I have just said." "I cannot do that," I replied. "Why not?" said the client. "Because I need to hear at least the rest of the sentence and, preferably, several more sentences to provide a meaningful interpretation," I responded. The man looked at me, bemused, then finished the sentence, which enabled me to do my job.

I hope my anecdote shows how dependent successful translation and interpretation are on context and the immediate linguistic, cultural, and physical environments. A somewhat meaningful interpretation of a partial statement may have been possible in that businessman's mother tongue due to its syntax and word-order rules. However, the rules of English syntax and the cultural context of the situation made such a partial interpretation impossible.

When the editors of this volume asked me to write this foreword, I welcomed the opportunity. As a former translator and interpreter, I find the volume's primary goal—to give technical communicators facility with translation and localization—very worthwhile. Also, as I read through the chapters, as a rhetorician, I deeply appreciated the book's attention to the rhetorical nature of translation and localization. As Bruce Maylath writes in the Preface, successful translators do not simply "look up the words" in a dictionary. Instead, successful translation and localization require both content creators and their translators to pay early and abundant attention to the main components of the rhetorical situation: purpose, audience, and context. Finally, the book's aim of combatting "the cult of monolingualism" is extremely important, especially in the current political and social climate.

The book is well-structured to fulfill its aims. The chapters are sequenced in a way that takes the readers from the more conceptual and theoretical to the practical. Each section of the volume tackles one or two important aspects of translation and localization work, giving the audience both a theoretical understanding of the problem at hand and a set of practical and tactical moves to solve it.

Many scholars and practitioners may see translation and localization as last steps of the communication process. However, several authors included in this volume make a case that technical communication and other content creators need to "front-load" their work with some of translation's and localization's concerns and challenges to make their projects more successful and effective with the end user.

Overall, then, at the risk of sounding more like a reviewer than the writer of a foreword, I have found this volume to be both useful and enjoyable to read. The chapters included in it are written in a clear and effective style and will be useful to anyone in professional communication and related fields. While the book emphasizes professional communication for business, commerce, science, and technology, it also implicitly argues for the importance of cultural understanding and ability to function in different environments and contexts. I appreciate both aspects of this book, both as a professional and a multilingual reader.

PREFACE

Handling Technical and Professional Communication for Audiences of All Languages

Bruce Maylath

"Translation just requires a secretary in the back room with a foreign language dictionary, doesn't it?" While interviewing translators and their clients for over two decades, I have found that this is a question that I hear again and again from professional communicators new to translation. The first I heard it was in 1997, when I spent the summer in residence at a translation company. The company's owner told me on the first day that this question was what so many of his first-time clients told him that they had asked themselves before finally, in frustration, turning to his company, after their secretaries couldn't make sense of what to do. Initially incredulous, I learned for myself the veracity of the company owner's claim when the owner put me in contact with a half dozen of the company's clients. Early in my interviews with them, client after client volunteered, nearly always with same words, "Yeah, before I started working with [translation company's name], I thought translations were just a matter of a secretary in the back room thumbing through a bilingual dictionary, substituting words."

Astonishingly, more than 20 years after the signing of the treaties creating the World Trade Organization and the North American Free Trade Agreement, the subsequent exponential increase in global trade, and consequent translation/localization projects, we still hear the question at the top of this page, both from new translation clients and their professional communicators inexperienced with translation projects, as well as from students in our technical and professional communication courses. The ignorance revealed in this question stems mainly from a cult of monolingualism, which I have described when addressing the IEEE Professional Communication Society (Maylath, 2016) and in subsequent presentations (Maylath 2017). It is reinforced by reactions against globalization—for example, the Trans-Pacific Partnership (TPP) having been condemned by the 2016 US presidential nominees of both major parties ("2016 Presidential Candidates" 2016), not long after the Brexit vote to prompt the United Kingdom's withdrawal from the European Union—and by comments by one of those presidential candidates bullying other primary contenders for their speaking to supporters in Spanish (Scott 2015). During this same period, a woman in the Minneapolis-St. Paul area severely injured a woman in a restaurant booth next to hers when she overheard fellow Americans conversing in Swahili. Using her beer mug as a club against her victim's face, the attacker exclaimed, "When you're in America you should speak English." The victim replied, "I can speak English, but we choose to speak whatever language we want" (Ibrahim 2015).

While the vast majority of professional communicators would never strike out in violence on hearing or seeing a language not their own, they do, however, render damage when they do not acknowledge their audience members' preferences for reading or listening to communications in their own languages. The cult of monolingualism easily leads to attitudes of dismissal and rejection of, if not outright hostility toward, other languages. Even when monolingual professional communicators are assigned a translation project for the first time and approach it with an attitude of curiosity and enthusiasm, myths and misconceptions often lead to costly, time-consuming errors, starting with the myth that translation requires nothing more than a secretary flipping through a dictionary but extending to the misconception that machine translation, including free services offered on the web, can provide a usable translation at little to no cost. With chapters contributed by a broad range of translators and instructors of translation, most of whom I have known and with whom I have worked for many years, this book aims to overcome such myths and misconceptions.

References

"2016 Presidential Candidates on the Trans-Pacific Partnership Trade Deal." 2016. *Ballotpedia.* https://ballotpedia.org/2016_presidential_candidates_on_the_Trans-Pacific_Partnership_trade_deal#tab=Democratic_ticket

Ibrahim, Mukhtar. 2015. "Applebee's Attack Victim: 'I'm Scared for My Life'." *MPR News.* November 6, 2015. www.mprnews.org/story/2015/11/06/applebees-attack-victim

Maylath, Bruce. 2016. "Examining the Cult of Monolingualism." Plenary address presented at the IEEE International Professional Communication Conference (ProComm), Austin, TX, October 3, 2016. Related interview available at https://procomm.ieee.org/bruce-maylath-on-the-cult-of-monolingualism/

Maylath, Bruce. 2017. "Where Has a Century of Monolingualism Got Us?" Lecture presented at the 56th Faculty Lectureship of North Dakota State University, Fargo, ND, April 6, 2017. www.youtube.com/watch?v=JHpCatOrmsA

Scott, Eugene. 2015. "Trump: Bush Should Speak English When in the U.S." *CNN.* September 3, 2015. www.cnn.com/2015/09/02/politics/donald-trump-jeb-bush-spanish/

INTRODUCTION

The Dynamics of—and Need to Understand—Translation and Localization in Technical Communication

Kirk St.Amant

The New Nature of International Interaction

In the last decade, international online access has grown almost exponentially. What was just over 360 million online connections in 2000 has blossomed into over 4.2 billion persons with internet access as of this writing ("Internet Growth Statistics" 2017, "Internet Usage Statistics" 2018). Such connectivity, moreover, ranges across a variety of media and forms. These include everything from static websites to dynamic social networking apps to online games comprised of players located around the world.

This global increase in online access has opened international markets on a scope and scale unthinkable a decade ago. India's online consumers, for example, represent roughly US$39 billion and could become an online consumer block of almost US$120 billion by 2020 (Press Trust of India 2016, Maheshwari 2016, Dalal 2016). In comparison, the online market in China is estimated at almost US$590 billion—and that number accounts for only 380 million of China's 1.4 billion citizens (Tong 2016, "China's E-Commerce Market" 2016). These developments make global markets relatively easy—and tempting—to tap if organizations can "speak the language" of this huge and growing base of potential customers.

The convergence of these factors brings with it an impetus to create materials for international audiences. The need for *translation*, for example, has become paramount as more organizations can now connect directly and (almost) instantly to individuals around the globe. For such access to be effective, materials need to be in the preferred (often the native) language and style of the prospective consumer (Ulijn and Strother 1995, Ulijn 1996). As the number of languages used online continues to grow, the need for quick and effective translation will only increase ("Top Languages of the Internet" 2015, "Internet World Users by Language" 2017). The implications for technical communicators—the professionals often charged with coordinating an organization's translation activities—are immense. But such factors encompass more than just translation.

Today's organizations also face an increasing need to develop materials that meet the design expectations and preferences of users from other nations. Doing so involves more than just language and often encompasses the configuration of products themselves (e.g., the visuals used on a website or the design of electrical plugs on an appliance). Addressing such issues involves more than providing materials in a format that audiences can recognize (e.g., I know what this *is*.). It also requires doing so in a manner that audiences consider acceptable (e.g., I consider this an *appropriate* way to address this topic) (Kostelnick 1995, St.Amant 2005). Failure to meet both conditions can result in cultural and legal repercussions affecting access to overseas markets (Kamath 2000). These factors mean the demand for *localization*—designing materials to meet the expectations of cultural audiences—is also on the rise. In fact, as of this writing, the global language industry (i.e., market for translation and localization services) is estimated to constitute US$40 billion in value, and it is the fourth fastest-growing industry in the United States (GALA 2017a).

Change and Contexts

For organizations wishing to compete in the new global economy, translation and localization are no longer ancillary practices reserved only for certain occasions. Rather, they have become core business practices central to "going global" (Rolink 2017, Agrwal 2017). In today's age of international online access, one could argue that the nature of cyberspace makes it almost impossible not to "go global" by virtue of going online (St.Amant and Rice 2015). This new context brings to the fore a myriad of variables that organizations need to address when tapping international markets. Effectively navigating such complexities requires an understanding of the elements affecting different interactions at multiple levels and in varying contexts (St.Amant 2015). The communication-based nature of such situations means that technical communicators are often at the heart of these activities. Moreover, these situations are now a reality affecting communication on a more local level.

Translation and localization have also become increasingly important at the local level. Shifting geopolitical trends mean that individuals are now moving internationally on an unprecedented scale. In the United States, for example, some 43.3 million immigrants were living in the country as of 2015—more than double the number of immigrants in the United States in 1990 ("Changing Patterns in U.S. Immigration and Population" 2014; Zong, Batalova, and Hallock 2018). In some cases, this international movement reflects economics —individuals from one nation moving to another in search of work (Milanovic 2013).

In other instances, these global migrations are associated with education and reflect students' seeking opportunities at educational institutions in other nations ("Choosing the USA" 2017). In yet other contexts, these shifts connect to hopes of safety and stability as refugees relocate to find experiences different from those in the lands that they left (Irons 2016).

These and other factors mean that the populations of nation states are increasingly changing in relation to the cultures that they contain, the languages spoken within their borders, and the communication and design needs of persons in their communities. For example, there are now more native Spanish speakers in the United States than in Spain—and many members of this audience prefer materials in Spanish, rather than in English (Romero 2017). Accordingly, the need to understand—and effectively employ—translation and localization services is as much a growing local need as it is a global imperative. These trends mean that state offices of motor vehicles need to understand the dynamics of language, culture, and communication almost as much as the multinational corporation operating in the same municipality. It is in this new global context that technical communicators are practicing their craft. Additionally, it is this new global context that they need to understand how to integrate successfully into the world economy of today and tomorrow.

The Nature of Technical Communication

Within these environments, technical communicators have become central to meeting the communication needs of cultural and linguistic groups. Many of the demands of the global economy—both international or local—involve creating effective, usable information (Sun and Getto 2017, St.Amant 2017a). Such information often takes the form of materials that explain how a technology works, how to perform an activity, or how to access a resource. This situation is based upon the overarching truth that technologies are not inherently usable. Rather, they reflect the expectations, preferences, and needs of the culture(s) that created them (Pacey 1996, Norman 2012, Sun 2012). As a result, one can never assume that an audience will instinctively know how to operate a new technology. Yet understanding how to use materials and engage in activities is essential to success in local and global settings. This makes the need to communicate such factors paramount to the success of any organization. And this area is the realm of the technical communicator.

It is technical communicators who have become the recognized experts at conveying information in ways that ensure an audience will use materials as expected. After all, technical communicators are the ones who develop the texts explaining how a process works or how a technology operates. They are also the professionals who craft the content needed to operate a device or perform processes to achieve a desired objective. What's more, technical communicators are the individuals generally charged with selecting the best mechanism or media for delivering content in easy-to-use ways.

Their positioning means that technical communicators often serve as an organization's central actors when communicating with the world. It is also why technical communicators have historically been tasked to coordinate with translators and localizers to create content for audiences from different cultural and linguistic backgrounds. These interactions have usually involved technical communicators doing everything from providing original source texts for translation to collaborating with overseas writers to produce parallel texts in other languages (Walmer 1999, Flint et al. 1999). Historically for localization, this relationship often involved technical communicators creating materials that localizers could more easily reconfigure to meet different cultural expectations (e.g., the design of an overall interface or website). In these situations, the connection was clear: technical communicators worked with translators and localizers to ensure that information moved effectively from one cultural and linguistic group to another.

The success (or failure) of such relationships was often connected to a central factor: understanding. The more technical communicators understood the translation and localization processes, the better they could design materials for them (Walmer 1999, Flint et al. 1999). Ideally, this understanding was based on knowing why communication practices in one's native culture needed to be revised for different cultural groups. In truth, such understanding often became encapsulated into "strategies." Such strategies generally consisted of quick, imperative steps for writing for translation or designing for localization (e.g., "Use active voice").

In some cases, these strategies attempted to convey a deeper understanding (e.g., "Avoid long noun strings, for translators have difficulty determining the subject of a noun string when translating texts"). The overall focus, however, often remained on *what* to do vs. *why* to write or design a certain way (e.g., "Avoid long noun strings"). The understanding needed to create content for translation and localization was often downplayed, lost, or overlooked in favor of quick tips for generating materials.

The Need to Know

In today's global economy, an understanding of other cultures is essential to an organization's success. This is because time has become compressed. Now, organizations are continually vying for international market share. To do so, they must create content that makes products readily usable on a global level (Sherwin 2016). They also need to address a growing number of national trade statutes and international commerce restrictions that make translating and localizing central to accessing overseas markets (Hunter 2016). Similarly, the need to create more kinds of content has increased, as has the need to create content for easy reuse to convey consistent messages across different media (e.g., web pages, Facebook posts, or Twitter tweets) (Knight 2017). These new contexts leave little room for error. This situation means that ineffective translations or localizations can spell the difference between capturing a key global market or missing out on an important international opportunity.

The same situation is occurring at the local level. The increasingly multilingual and multicultural settings within nations mean that successfully tapping local markets increasingly involves translation and localization (Romero 2017). Additionally, changing national laws are increasingly affecting local practices. Statutes, policies, and mandates, like US Executive Order 13166, can require state and local agencies to translate and/or localize materials or face penalties (St. Germaine-McDaniel 2010). Such factors contribute to an increased need for effective translation and localization in a growing number of local contexts. Their importance has led to the publication of academic journal issues dedicated to examining translation and localization within technical communication. These include Bruce Maylath, Ricardo Muñoz Martín, and Marta Pacheco Pinto's 2015 issue of *connexions: International Professional Communication Journal* focusing on "Translation and International Professional Communication" and Huatong Sun and Guiseppe Getto's 2017 special issue of *Technical Communication* focusing on "Localizing User Experience."

These trends point to one central reality of the modern economy: Technical communicators need to understand translation and localization to contribute value to organizations ("Body of Knowledge" 2017). The first step toward such an understanding is *realization*. Specifically, technical communicators need to *realize* the connections between technical communication, translation, and localization. They then need to use this knowledge to build on these connections in meaningful ways that foster understanding. Doing so requires all parties to recognize that these different professions all focus on a common objective: conveying meaning effectively from one group to another.

Connections and Differences

Technical communication is often likened to translation. In both fields, trained individuals are responsible for conveying meaning from one group to another (Puchleitner and Petrovic 2014, Center for Environmental Excellence 2014). Such transfers of meaning often involve bridging different systems for conveying ideas—be they from mathematical equation to printed text or from one language system to another. Given such similarity, one might assume that technical communicators are well equipped to work with translation and localization professionals. Unfortunately, this situation is not always—or even often—the case.

As prior technical communication research notes, the process of translation involves more than converting individual words from one language to another.

Rather, it involves understanding the nuances and complexities of conveying meaning—what those words, in the aggregate, attempt to convey—across cultural spheres (GALA 2017b). Such transfer encompasses more than language and words. It also involves style, structure, and nuance—or rhetoric—when sharing information with different cultural groups. In some cases, these differences can include where to present a particular kind of information in a document (Driskill 1996, Tebeaux 1999). In others, it can involve how much and what kind of information that the members of a culture expect to encounter in a communique (Tebeaux 1999, St.Amant 2006). It can also mean determining which genres/kinds of communiques to use when conveying an idea to other cultures or if certain genres even exist in other cultures (Woolever 2001, St.Amant 2006). What's more, it might require the creation of new terms as a parallel word or phrase for a concept might not exist in another culture or language (Mikelonis 2000).

At the same time, new contexts and technologies are affecting how individuals view translation within technical communication (Gonzales 2018). The global diffusion of multimedia communication technologies, for example, has prompted inquiries into whether translation should include non-verbal factors, such as facial expressions, gestures, and posture, when addressing video-based messages (Gonzales and Zantjer 2015). Similarly, the growing importance of translation in specialty areas like medical communication has prompted work on how to convey health-related information across languages and cultures (Zhu and St.Amant 2007, Batova 2010). Then there is the growing interest in how to employ approaches from usability and user experience design (UXD) to create products for other cultures (Sun 2012, Getto and St.Amant 2014, Sun and Getto 2017; St.Amant 2017a).

Folded into these practices are ideas of *glocalization* and *transcreation*: creating completely new, culture-specific versions of a product. The idea is such versions are specifically designed to meet the expectations of the cultures that will use them, in contrast to more conventional localization practices in which products were created for one culture and reconfigured or redesigned to meet the expectations of another (Merino 2006, Sun 2012, Pedersen 2014, Munday 2016). The resulting situation is complex, to say the least. For technical communicators new to such processes, this situation can seem truly overwhelming.

Practices in Pedagogy

These developments are also affecting technical communication education. Now the topic of working in international context—with a particular focus on translation and localization—appears in a number of introductory textbooks in the field. John M. Lannon and Laura J. Gurak's (2017) *Technical Communication, 4th edition*, for example, contains dedicated sections on "Identifying Your Audience's Cultural Background," "Global Considerations When Working in Teams," and "Considering the Global Context." A similar situation is true for instructional texts that focus on more specialized areas in technical communication. Carolyn Rude and Angela Eaton's (2010) textbook *Technical Editing, 5th edition*, for example, has an entire chapter dedicated to aspects of translation and localization. Likewise, Barbara A. Heifferon's (2005) *Writing in the Health Professions*—a textbook targeting a segment of the technical communication field—includes an entire chapter on "Multicultural and International Medical Writing."

This focus on translation and global issues is mirrored in the pedagogical research of the field. Bruce Maylath's (1997) seminal *JBTC* article "Teaching the Technical Writing Student to Prepare Documents for Translation" marks one of the first examinations of this area of instruction. Educators have since expanded upon these early examinations to mirror the changing nature of technical communication pedagogy. The focus on culture and communication in technical communication education has been integrated into the context of online education (St.Amant 2017b, Slattery and Cleary 2016), which has also been studied in how to teach students to work in globally distributed virtual teams (Brewer 2016, Cleary et al. 2015). More recently, the teaching of translation has been explored in relation to service learning and community outreach and engagement projects (Gonzales 2017).

So central is international understanding to technical communication education that it represents a core program development area ("Interculturality") in the TecCOMFrame project for harmonizing technical communication curricula in Europe ("Academic Competence Framework," n.d.). Thus, the field's pedagogical practices are aligning with those of industry. Now both recognize the importance of working effectively with translators and localizers to succeed in the modern global workplace.

Recognition and Understanding

Addressing such factors successfully involves developing an effective knowledge of the translation and localization fields. Gaining such knowledge is increasingly essential to the technical communicator's role in an organization. This situation also represents a key area where technical communicators can add value, in large and very public ways, to their organizations (Giammona 2004, "Body of Knowledge" 2017). In this context, the question becomes, "What to do and where to start?" The answers can be found in addressing two interrelated areas: recognition and understanding.

Recognition is an internal process that involves realizing the limits of one's knowledge and abilities. As a field of professionals, technical communicators increasingly encounter this idea of recognition via work noting the need to better understand the nuances of the modern global context (Giammona 2004, St.Amant and Melonçon 2016). As noted, there is widespread recognition that technical communicators need to address issues of translation and localization as a part of their jobs. *Understanding*, by contrast, involves the external. It requires individuals to move outside of the realm of what they know and to seek out knowledge that is new and possibly different. The need now is for resources that help technical communicators better grasp the dynamics of the translation and localization professions. This collection is such a text.

The nine entries in this volume introduce technical communicators to foundational practices, current approaches, and emerging trends in translation and localization. Central to the contribution that these chapters make is the authors who wrote them. Past technical communication scholarship in these areas was often limited by the researchers' abilities to understand "other" professions. That is, technical communicators were often studying or observing translation and localization practices and then reporting their experiences and observations (see, for example, Walmer 1999 and Mikelonis 2000).[1] Without being a part of and knowing the fields that they studied, most of these researchers were limited in their abilities to capture and analyze the nuances of the contexts that they examined. The level of understanding that such materials provided on translation and localization practices was thus limited. The entries in this volume seek to overcome these constraints.

To do so, the editors of this collection purposefully recruited authors from the translation and localization industries. All of these contributors, moreover, have previous experiences, stretching back decades, working with technical communicators on international projects. Each individual was asked to contribute a chapter that explained, to a non-translator/non-localizer, what a particular translation or localization practice was and how it worked. Contributors were also asked to draw from their prior experiences working with technical communicators and focus on the question, "Based on your prior experiences working with technical communicators, what do *they* need to know to work effectively with translators and localizers?"

The objective underlying this approach was to provide technical communicators with the understanding of translation and localization needed to work effectively with professionals in these fields. The resulting chapters form this text. To this end, the book that you are holding is organized into three major sections based on an examination of various key aspects of these professions.

The book's first section, Translation, centers on the practices, approaches, and ideas central to the translation profession. The second section, Localization, examines core processes and various trends that technical communicators need to understand to work effectively with localization specialists (many of whom are, in fact, translators). The third and final section, Horizons, reviews how shifts in technology and economics are affecting the "language industry" and "language service providers," general terms often used to encompass both translation and localization practices. In so doing, the chapters in this final section outline dynamics affecting practices in both translation and localization. They also reveal the overlap and close relationship between translation and localization and how fluid the dynamics across these two professional areas are.

Through this organizational structure, technical communicators can gain a fundamental understanding of practices and approaches in and across translation and localization. They can also gain insights into trends that will influence the development of these fields in the future. Moreover, these entries can provide technical communicators with a foundational knowledge of factors that can affect work done in both areas. With these ideas in mind, let's look at the sections and the entries in this collection.

Organization of the Text

The first major section, "Translation," provides an overview of the dynamics of moving meaning from one language and culture to others. The contributing authors review factors influencing and affecting translation practices. To do so, this overall section is split into two "Areas" that move readers from understanding more fundamental practices to techniques for generating content for translation.

The initial Area in the Translation section focuses on "central concepts" that guide practices in the translation industry. The section's first chapter, Federica Scarpa's "An Overview of the Main Issues of Translation," introduces readers to what translation is, how its core practices operate, and the basic problems that can affect translation work. The second chapter, "Pragmatic Translation and Assessment for Technical Communicators: Countering Myths and Misconceptions," by Patricia Minacori, builds upon Scarpa's chapter to examine the methods that translators use when moving meaning from one language to another. The combined information in these initial chapters provides technical communicators with a foundational understanding of what translators do when working with texts, how, and why. Technical communicators can use this information to consider how their actions as content creators and editors can hinder or help the work translators do.

The second Area of examination in the Translation section is "Writing and Content Creation." The entries in this section review how the processes that technical communicators use to create text can affect the ways in which translators approach a project. The first entry in this Area, Sharon O'Brien's "Controlled Language and Writing for an International Audience," offers technical communicators suggestions for how to craft texts to make them easier to translate efficiently and effectively. The next entry, "Quality in Translation through 'Controlled' Writing, Editing, and/or Revising," by Maria Teresa Musacchio, reviews the connections between the quality of the translation products created and the methods used to generate original source texts. Through their examination of these topics, the entries in this Area build on the concepts overviewed in earlier chapters. They also provide technical communicators with effective ideas for how to use that knowledge to work more effectively with translators.

The volume's second major section, "Localization," focuses on the practices within and the forces affecting this segment of the language industry. Unlike the discussion in the Translation section, this section organizes entries into one thematic "Area" that examines "Fundamentals, Contexts, and Content" essential to understanding and working with members of this profession. The first chapter here, Paola Valli's "Fundamentals of Localization for Non-Localizers," is an introduction to what localization is. The author also reviews how localization is connected to translation and the overall language industry and how organizations view localization services.

The next two entries in this Area examine how the setting where localization occurs affects the dynamics under which technical writers create source materials for localization. The first of these two chapters, Birthe Mousten's "How Streetsmart Marketers Try Phishing Customers: (Mis)Communication or Missed Communication through Loan Words," reviews how international business practices have affected approaches to localization. In the next chapter, "Technicial Communicators on Track or Led Astray! Using Quality Standards in Practice," Mousten partners with the late Dan Riordan to review how technical communicators can use virtual teams to collaborate with localizers. In examining these topics, the two entries review different contexts and methods that technical communicators can use to work with localizers to generate source materials. Accordingly, the chapters in this section create a starting point from which technical communicators can explore ways to engage localizers as partners in international content creation.

The book's third and final major section, "Horizons," examines factors that overlap both translation and localization. The thematic "Area" for this section focuses on how factors of technology ("Computing") and economy ("Commerce") are changing the translation and localization industries. In so doing, these entries delineate how such forces might shape these fields in years to come. For technical communicators, such information could be instrumental to maintaining an up-to-date understanding of the translation and localization professions over time.

The first chapter in this section, Tomáš Svoboda's "Computing and Translation: An Overview for Technical Communicators," reviews how computing technologies, such as translation software and translation memory, are changing the work that translators do. In reviewing these topics, Svoboda also reveals the changing contexts that technical communicators must consider when they create source texts for translators. The section's second and final entry, "The Future of Translation and Translators in a Fast-Changing Economic and Technological Landscape," by Giuseppe Palumbo, reviews the convergence of global business practices affecting translation and localization. Palumbo also reveals how advances in computing technology are changing the ways in which translators, and the organizations that employ them, conceptualize communication practices. By examining these issues, the authors of these chapters introduce technical communicators to the technological and economic forces shaping the modern language industry. These entries also provide technical communicators with frameworks for identifying and understanding such factors. In so doing, these chapters provide readers with approaches to addressing such dynamics and working effectively with translators and localizers over time.

Concluding Thoughts

A map is an essential tool for navigating a new territory. But a map is only a tool; how it is used depends on the individual holding it. In many ways, this text represents a map that technical communicators can use to navigate the relatively unknown areas of translation and localization. Like a map, this text provides markers that one can use to gauge progress on his or her travels. The various sections and areas of this book are designed as just that: markers. As technical communicators move through the text, they move deeper and deeper into their understanding of what translators and localizers do, how they do it, and why.

This guide, moreover, can be used by individuals from across technical communication. First, it can help industry practitioners work more successfully with translators and localizers to create effective products for greater global markets. Second, it can provide academic researchers with the level of understanding needed to investigate the dynamics of cross-profession collaboration in informed and meaningful ways that benefit all involved. Third, it can provide technical communication educators with the understanding essential to helping students develop the skills and knowledge needed to work with translation and localization professionals upon graduation.

Like all maps, the background of the reader is important. However, the fact that all readers are navigating toward a common goal—an increased understanding of translation and localization practices—allows this book to have value across different groups within the field. In fact, this collection could foster closer relations by providing practitioners, researchers, and educators with a common foundation for collaboration around a mutual interest. In these ways, this text can help technical communicators negotiate the landscape of new and different fields and navigate the various areas of their own discipline.

Note

1 In certain cases, such as Flint et al. 1999, a trained translator was involved in the research process, but this involvement was often the exception.

References

"Academic Competence Framework, The." n.d. *TecCOMFRame*. www.teccom-frame.eu/competence-framework/overview/

Agrwal, A. J. 2017. "Ten Reasons Why Businesses Are Going Global." *The Huffington Post*. August 14, 2016. www.huffingtonpost.com/aj-agrawal/ten-reasons-why-businesse_b_11512636.html

Batova, Tatiana. 2010. "Writing for the Participants of International Clinical Trials: Law, Ethics, and Culture." *Technical Communication* 57, no. 3: 266–281.

"Body of Knowledge." 2017. *Society for Technical Communication*. www.tcbok.org/wiki/about-tc/value-proposition/

Brewer, Pam Estes. 2016. "Teaching and Training with a Flexible Module for Global Virtual Teams." In *Teaching and Training for Global Engineering: Perspectives on Culture and Professional Communication Practices*, edited by Kirk St.Amant and Madelyn Flammia, 173–197. Hoboken, NJ: Wiley-IEEE Press.

Center for Environmental Excellence. 2014. *Preparing High-Quality NEPA Documents for Transportation Projects*. Washington, DC: Center for Environmental Excellence. www.environment.transportation.org/pdf/programs/pg15-1.pdf

"Changing Patterns in U.S. Immigration and Population: Immigrants Slow Population Decline in Many Counties. 2014. *Pew Charitable Trust*. www.pewtrusts.org/en/research-and-analysis/issue-briefs/2014/12/changing-patterns-in-us-immigration-and-population

"China's E-Commerce Market: Untapped Potential for Global Companies." 2016. *Nielsen*. http://sites.nielsen.com/newscenter/chinas-e-commerce-market-untapped-potential-for-global-companies/

"Choosing the USA." 2017. *International Student*. www.internationalstudent.com/study_usa/choosing-the-usa/

Cleary, Yvonne, Madelyn Flammia, Patricia Minacori, and Darina M. Slattery. 2015. "Global Virtual Teams Create and Translate Technical Documentation: Communication Strategies, Challenges and Recommendations." In *Proceedings of the 2015 IEEE Professional Communication Conference, Limerick, Ireland, July 12–15*, 1–10. Piscataway, NJ: IEEE.

Dalal, Mihir. 2016. "So, What's the State of India's Online Retail Market Anyway?" *Live Mint*. www.livemint.com/Companies/Ig34ixeE1nJZAjMSl5Wg0O/So-what-is-the-state-of-Indias-online-retail-market-anyway.html

Driskill, Linda. 1996. "Collaborating Across National and Cultural Borders." In *International Dimensions of Technical Communication*, edited by Deborah C. Andrews, 23–44. Alexandria, VA: Society for Technical Communication.

Flint, Patricia, Melanie Lord van Slyke, Doreen Stärke-Meyerring, and Aimee Thompson. 1999. "Going Online: Helping Technical Communicators Help Translators." *Technical Communication* 46, no. 2: 238–248.

Getto, Guiseppe, and Kirk St.Amant. 2014. "Designing Globally, Working Locally: Using Personas to Develop Online Communication Products for International Users." *Communication Design Quarterly* 3, no. 1: 24–46.

Giammona, Barbara. 2004. "The Future of Technical Communication: How Innovation, Technology, Information Management, and Other Forces Are Shaping the Future of the Profession." *Technical Communication* 51, no. 3: 349–366.

GALA (Globalization and Localization Association). 2017a. "Translation and Localization Industry Facts and Data." *GALA*. www.gala-global.org/industry/industry-facts-and-data

———. 2017b. "What Is Translation?" *GALA*. www.gala-global.org/industry/intro-language-industry/what-translation

Gonzales, Laura. 2018. *Sites of Translation: What Multilinguals Can Teach Us About Digital Writing and Rhetoric*. Ann Arbor: University of Michigan Press.

———. 2017. "But Is That Relevant Here? A Pedagogical Model for Embedding Translation Training Within Technical Communication Courses in the US." *connexions:*

International Professional Communication Journal 5, no. 1: 75–108. https://connexionsj.files.wordpress.com/2017/06/gonzales2.pdf
Gonzales, Laura, and Rebecca Zantjer. 2015. "Translation as a User-Localization Practice." *Technical Communication* 62, no. 4: 271–284.
Heifferon, Barbara A. 2005. *Writing in the Health Professions.* New York: Allyn & Bacon.
Hunter, Mike. 2016. "The Role of Translation When Entering International Markets." *Open to Export.* http://opentoexport.com/article/the-role-of-translation-when-entering-international-markets/
"Internet Growth Statistics. 2017. *Internet World Stats.* www.internetworldstats.com/emarketing.htm
"Internet Usage Statistics." 2018. *Internet World Stats.* www.internetworldstats.com/stats.htm
"Internet World Users by Language." 2017. *Internet World Stats.* www.internetworldstats.com/stats7.htm
Irons, Jeannie. 2016. "Crossing the Border: Why Refugees Flee Their Homes." *Children International.* November 8, 2016. www.children.org/stories/2016/november/why-refugees-flee-their-homes
Kamath, G. R. 2000. "The India Paradox." *Intercom*, May, 10–11.
Knight, Warren. 2017. "How to Create a Social Media Marketing Content Plan in 7 Steps." *Social Media Examiner.* April 27, 2017. www.socialmediaexaminer.com/how-to-create-social-media-marketing-content-plan-in-7-steps/
Kostelnick, Charles. 1995. "Cultural Adaptation and Information Design: Two Contrasting Views." *IEEE Transactions on Professional Communication* 38, no. 4: 182–196.
Lannon, John M., and Laura J. Gurak. 2017. *Technical Communication.* 14th ed. New York: Pearson.
Maheshwari, Richa. 2016. "Indian Ecommerce Market to Grow Fastest Over 3 Years." *The Economic Times.* February 18, 2016. http://economictimes.indiatimes.com/industry/services/retail/indian-ecommerce-market-to-grow-fastest-globally-over-3-years-morgan-stanley/articleshow/51031652.cms
Maylath, Bruce. 1997. "Teaching the Technical Writing Student to Prepare Documents for Translation. *Journal of Business and Technical Communication* 11, no. 3: 339–352.
Maylath, Bruce, Ricardo Muñoz Martín, and Marta Pacheco Pinto. 2015. "Translation and International Professional Communication: Building Bridges and Strengthening Skills." *connexions: International Professional Communication Journal* 3, no. 2: 3–9. https://connexionsj.files.wordpress.com/2016/02/maylath-munoz-pinto.pdf
Merino, Miguel Bernal. 2006. "On the Translation of Video Games." *The Journal of Specialized Translation* 6: 22–36. www.jostrans.org/issue06/art_bernal.php
Mikelonis, Victoria M. 2000. "Message Sent Versus Message Received: Implications for Designing Training Materials for Central and Eastern Europeans." In *Managing Global Communication in Science and Technology*, edited by Peter J. Hager and H. J. Scheiber, 203–231. New York: John Wiley & Sons.
Milanovic, Branko. 2013. "The Economic Causes of Migration." *The Globalist.* October 22, 2013. www.theglobalist.com/economic-causes-migration/
Munday, Jeremy. 2016. *Introducing Translation Studies: Theories and Applications.* 4th ed. New York: Routledge.
Norman, Don. 2012. "Does Culture Matter for Product Design? *Core 77.* January 9, 2012. www.core77.com/posts/21455/Does-Culture-Matter-for-Product-Design
Pacey, Arnold. 1996. *The Culture of Technology.* Cambridge, MA: MIT Press.

Pedersen, Daniel. 2014. "Exploring the Concept of Transcreation: Transcreation as 'More Than Translation'?" *Cultus: The Journal of Intercultural Mediation and Communication*, 7: 57–71.

Press Trust of India. "India's E-Commerce Industry Likely to Touch $38 Billion Mark in 2016." 2016. *YourStory*. January 2, 2016. https://yourstory.com/2016/01/indias-e-commerce-assocham/

Puchleitner, Thomas, and Otto Petrovic. 2014. "From Manuals Towards Product Embedded Interactive Learning Environments." In *2014 Global Engineering Education Conference (EDUCON), Istanbul, Turkey, April 3–5*, 243–248. Piscataway, NJ: IEEE.

Rolink, Jacek. 2017. "American Companies Expanding Overseas: Trend or Necessity." *Blueback Global*. May 22, 2017. https://bluebackglobal.com/business/american-companies-expanding-overseas/

Romero, S. 2017. Spanish Thrives in the U.S. Despite an English-Only Drive. *The New York Times Online Edition*. August 23, 2017. www.nytimes.com/2017/08/23/us/spanish-language-united-states.html?emc=eta1

Rude, Carolyn D., and Angela Eaton. 2010. *Technical Editing*. 5th ed. New York: Allyn & Bacon.

Sherwin, Katie. 2016. "Cultural Nuances Impact User Experience: Why We Test with International Audiences." *Nielsen Norman Group*. June 26, 2016. www.nngroup.com/articles/cultural-nuances/

Slattery, Darina M., and Yvonne Cleary. 2016. "E-Learning and Technical Communication for International Audiences." In *Teaching and Training for Global Engineering: Perspectives on Culture and Professional Communication Practices*, edited by Kirk St.Amant and Madelyn Flammia, 149–172. Piscataway, NJ: Wiley-IEEE Press.

St.Amant, Kirk. 2005. "A Prototype Theory Approach to International Web Site Analysis and Design." *Technical Communication Quarterly* 14, no. 1: 73–91.

———. 2006. "Globalizing Rhetoric: Using Rhetorical Concepts to Identify and Analyze Cultural Expectations Related to Genres." *Hermes: Journal of Language and Communication Studies* 37: 47–66.

———. 2015. "Cultural Considerations for Communication Design: Integrating Ideas of Culture, Communication, and Context into User Experience Design." *Communication Design Quarterly* 4, no. 1: 6–22.

———. 2017a. "The Cultural Context of Care in International Communication Design: A Heuristic for Addressing Usability in International Health and Medical Communication." *Communication Design Quarterly* 5, no. 2: 62–70.

———. 2017b. "Of Friction Points and Infrastructures: Re-Thinking the Dynamics of Offering Online Education in Technical Communication in Global Contexts." *Technical Communication Quarterly* 26, no. 3: 1–19.

St.Amant, Kirk, and Lisa Melonçon. 2016. "Reflections on Research: Examining Practitioner Perspectives on the State of Research in Technical Communication." *Technical Communication* 63, no. 4: 346–363.

St.Amant, Kirk, and Rich Rice. 2015. "Online Writing in Global Contexts: Rethinking the Nature of Connections and Communication in the Age of International Online Media." *Computers and Composition*, 38(B): v–x.

St. Germaine-McDaniel, Nicole. 2010. "Technical Communication in the Health Fields: Executive Order 13166 and Its Impact on Translation and Localization." *Technical Communication* 57, no. 3: 251–265.

Sun, Huatong. 2012. Cross-Cultural Technology Design: Creating Culture-Sensitive Technology for Local Users. New York: Oxford University Press.

Sun, Huatong, and Guiseppe Getto. 2017. "Localizing User Experience: Strategies, Practices, and Techniques for Culturally Sensitive Design." *Technical Communication* 64, no. 2: 89–94.

Tebeaux, Elizabeth. 1999. "Designing Written Business Communication Along the Shifting Cultural Continuum: The New Face of Mexico." *Journal of Business and Technical Communication* 13: 49–85.

Tong, Frank. 2016. "China's Online Retail Sales Grow a Third to $589 Billion in 2015." *Internet Retailer.* January 27, 2016. www.digitalcommerce360.com/2016/01/27/chinas-online-retail-sales-grow-third-589-billion-2015/

"Top Ten Languages of the Internet, Today and Tomorrow." 2015. *Unbabel.* June 10, 2015. https://unbabel.com/blog/top-languages-of-the-internet/

Ulijn, Jan M. 1996. "Translating the Culture of Technical Documents: Some Experimental Evidence." In *International Dimensions of Technical Communication*, edited by Deborah C. Andrews, 69–86. Arlington, VA: Society for Technical Communication.

Ulijn, Jan M., and Judith B. Strother. 1995. *Communicating in Business and Technology: From Psycholinguistic Theory to International Practice.* Frankfurt: Peter Lang.

Walmer, Daphne. 1999. "One Company's Efforts to Improve Translation and Localization." *Technical Communication* 46, no. 2: 230–237.

Woolever, Kristin R. 2001. "Doing Global Business in the Information Age: Rhetorical Contrasts in the Business and Technical Professions." In *Contrastive Rhetoric Revisited and Redefined*, edited by Clayann Gilliam Panetta, 47–64. Mahwah, NJ: Lawrence Erlbaum Associates.

Zhu, Pinfan, and Kirk St.Amant. 2007. "Taking Traditional Chinese Medicine International and Online: An Examination of the Cultural Rhetorical Factors Affecting American Perceptions of Chinese-Created Web Sites." *Technical Communication* 54, no. 2: 171–186.

Zong, Jie, Jeanne Batalova, and Jeffrey Hallock. 2018. "Frequently Requested Statistics on Immigrants and Immigration in the United States." *Migration Policy Institute.* February 8, 2018. www.migrationpolicy.org/article/frequently-requested-statistics-immigrants-and-immigration-united-states

PART I
Translation

AREA 1

CENTRAL CONCEPTS

1

AN OVERVIEW OF THE MAIN ISSUES OF TRANSLATION

Federica Scarpa

Introduction

Translation helps transcend language barriers and interact with people in different countries. The word "translation," however, is a general label for many activities and products. Broadly speaking, "translation," in the words of Bellos (2011) is "a two-headed beast"; without an article ("a[n]," "the"), it names "the process by which 'a translation' comes to exist," but with an article, it names "a product—any work translated from some other language" (21). "Translation" thus refers to

- the *process* of transferring meaning from an original text written in a source language to another language according to the specific socio-cultural context of that language and
- the *product* resulting from that process: the target (i.e., final, translated) text that should address the socio-cultural context of the intended audience reading in the target language.

The word "translation" can also refer to other activities and products based on criteria such as the following:

- Medium: Written, oral, audiovisual, etc.
- Mode: Conversion of a text from one language to another including
  - *intra*lingual translation: within the same language,
  - *inter*lingual translation: between different languages, and
  - intersemiotic translation: between different verbal/non-verbal systems such as from a novel to the medium of film

- Direction: Conversion of a text from or into the translator's mother tongue or language of habitual use
- Tools: Technologies used to help with translation (e.g., computer-assisted translation, machine translation, terminology-management systems etc.)
- Type: Kind of text to be translated (e.g., literary, legal, technical, political, etc., and, within such categories, genres such as journal article, science textbook, newspaper editorial, etc.)
- Completeness: Degree of completeness of the finished translation (e.g., translation of an entire text, parts of a text, summaries of a text, etc.)

These are aspects that technical communicators need to consider when preparing texts for translation.

This chapter focuses on *technical translation*,[1] a sub-type of specialized translation involving "technology based" texts—or texts produced by technical communicators and relating to technological subjects. In this case, technical information is translated to share it in a different language and with a new audience. To achieve this objective, information should be easily accessible in terms of comprehensibility, clarity, and speed of delivery (see Byrne 2006, 11).

By overviewing central concepts, processes, and practices in translation, this chapter can help acquaint technical communicators with the main issues of translation and the factors and constraints affecting it. To help technical communicators better understand such practices, the chapter has been organized into the following sections:

- "What is Technical Translation?" explains technical, scientific, and specialized translation for individuals new to such topics.
- "The Process of Translation" describes translation through three main stages of the translation process.
- "Translating Technical Texts: Problems and Strategies" explores common problems encountered by translators and the strategies often used to address them effectively.
- "Quality in Technical Translation" presents parameters for evaluating the quality of a technical translation, and it describes the translation process in terms of a translation workflow.
- "Training and Accreditation of Translators" describes training translators to achieve the needed mix of competences and skills.

Each section thus introduces technical communicators to the key translation processes, practices, and approaches and provides them with the understanding essential to working more effectively with translators and translation.

What Is Technical Translation?

In technical translation, the medium involved is usually written text, and translators are typically asked to translate whole documents, not just portions of them. The direction of the translation (i.e., the movement of a text from one language to another) is generally from a second language the translator learned (L2) into the translator's native language or language of habitual use (L1). The idea of individuals translating into their L1 is important, for one knows the nuances of her or his L1 needed to avoid "interference"—that is, repeating language patterns that are atypical of the target language (e.g., selecting the best idiomatic expression for conveying an idea). An example of such a situation in English is the expression *social dinner*—the result of interference from the Italian noun phrase *cena sociale*. The term needed to convey the same idea to English users is "conference dinner."

Despite the traditional rule that translators should always translate into their L1, translating into the translator's L2 is becoming increasingly common as a result of the growing use of technologies in the translation process. Such situations are particularly the case when the translator has a very specialized subject-matter expertise and the translation is revised by a native speaker (e.g., having a reviser or, indeed, the translator herself review marketing materials translated into that translator's L2 via a machine/software). Such a scenario has become much more acceptable thanks to the introduction of translation technology (see Svoboda's chapter in this volume).

Style and Register

In technical translation, the texts to be translated can be broadly categorized according to their subject matter as "technology based." Taken on its own, however, subject matter is not sufficient to help translators produce target-language texts that are virtually identical to, and therefore as informative as, texts produced by technical writers in that same language. To do this, they also need to acquire the relevant "style," loosely defined by Byrne (2006) as "the way we write things, the words we choose and the way we construct sentences" (4).

Language use is governed by different norms and conventions in different situations requiring different author/addressee relationships (peer-to-peer, popularizing, promotional, instructive, etc.). This relationship between language and the immediate situation in which it will be used means that technical translators need to acquire the different "registers" (e.g., formal, colloquial, institutional, etc.) in the target language in order to "package" textual material along patterns familiar to the target reader (Baker 2011, 123). For example, nobody would use the same register to describe a device or tool in both a newspaper entry and an article in a professional journal. So, translators must know

- the subject matter of the text to be translated,
- the context in which the translated text will appear (e.g., journal article, instruction manual, textbook, report, etc.), and
- the translation's main communicative purpose (expository, argumentative, instructional, etc.).

This information is useful in helping translators understand the style and the specific register of the technical language used in both the source (original) and the target (translated) text. This knowledge enables translators to choose the translation strategies needed to overcome translation problems they might encounter during their task (Hatim and Munday 2004, 74).

Technical vs. Scientific vs. Specialized Translation

Technical translation is not the same as scientific translation, for "science produces ideas about how the world works, whereas the ideas in technology result in usable objects" (Wolpert 2002, 969). Scientific texts deal with pure scientific knowledge; their aims are to describe, classify, discuss, explain, conclude, and justify. Technological texts focus on subjects based on *applied* knowledge from the natural sciences. The main function of such technological texts is utilitarian: to help someone do something.

Writing about science and technology can vary in terms of subject matter, purpose, and type of language. In a standard situation of peer-to-peer writing, the language of science is quite formal, abstract, and conceptually difficult, because it packs information into short phrases. Conversely, the language of technology is generally more concrete, referring, as it does, to products and processes in the external world. It is also more colloquial and conceptually easier, because it contains less information in more space (Byrne 2006, 7–10). An example illustrating such differences is "glass crack growth rate is associated with applied stress magnitude" vs. "glass cracks more quickly the harder you press it" (Halliday 2004, 196). The condensed meaning of the highly structured and abstract scientific language of the first statement has been "unpacked" into the easier and more concrete standard language of the second.

Finally, the collective expression *specialized translation* is used for technical, scientific, and legal-administrative translation (i.e., legal text used within businesses, organizations, and governments). Specialized translation comprises most of the world's annual translation output, and it is by far the most significant employer of translators (Byrne 2006, 2). According to recent data, in the European Union the highest demand for translation services is concentrated in the legal, life sciences, and financial content areas, with high-tech industries ranking lower (LTC 2016, 1). The lower ranking of technical translation, however, somewhat clashes with Olohan's (2016) claim that technical translation is an "activity [that] is flourishing in today's global economy and information society, in which there is strong demand for product specifications, instruction leaflets, user guides, etc., in many languages, as well as for the localization of software applications" (246). Even more strikingly, Byrne (2006, 2) estimated that technical translation, including only technological texts (i.e., excluding legal, business, financial, etc. texts), accounted for as much as 90% of the world's total translation output each year.

The Process of Translation

The process of translation refers to the cumulative process involving "what happens linguistically and cognitively as the translator works on the translation" (Hatim and Munday 2004, 346). Three basic stages of the translation activity lead to the translated text. These stages are

1. analysis of the source-language text,
2. transfer of the source text into the target language, and
3. revision of the translation.

This section examines each of these stages in more detail so that technical communicators can better understand each stage of the overall translation process.

Analysis of the Source-Language Text

In the first stage, the translator analyzes the source-language text (meaning, register, terminology, etc.) for two reasons. The first is to gain a complete understanding of the text to be translated. The second is to select a general approach that will guide the translator's choices at the various levels of the text. Achieving this second goal depends on gaining an accurate understanding of the purpose of the translation and the needs of the target readers (acquired from the so-called "translation brief"—see Durban 2011).

The reading that the translator does at this stage needs to be extremely careful. Unlike when other individuals read the same text, translators identify weak spots in the original text—areas where clarification is needed. This process allows one to improve the text *even in the source language*. This step helps translators serve as technical communicators' best reviewers and proofreaders.

Transfer into the Target Language

The second stage involves the transfer or reformulation of the ideas from the source text into the target language. At this stage, trained translators solve translation problems that they encounter by applying appropriate strategies that they have been taught. Such problems refer to the translatability of the source text and typically include the following:

1. Linguistic problems relating to the wording of the specific text to be translated, such as terminological problems linked to the translator's (lack of) knowledge of the specific technical terminology used in the text
2. Pragmatic and cultural problems linked to the different audience and/or communicative purpose of the original and translated texts, as well as the different norms and conventions governing language use in the same

situation in different languages. Examples range from the date format (in the United States "07.12.18" indicates 12 July 2018 and not 7 December 2018, as in Italian) to the problem of translating the meteorological term *wintry shower* in a target culture lacking such a concept, which can be solved by explaining that term in the target language (e.g., "brief but intense snow or sleet precipitation") (Scarpa 2002).

Through this process, translators produce a draft translation using all the tools available for researching and acquiring relevant information. These tools include terminological databases, specialized documents, discussion forums, translation memories, etc. As with all text processing, translation too is largely a matter of problem-solving. This situation means the fewer problems in the source text to be translated, the faster, more effective, and cheaper the resulting translation will be.

Revision of the Translation

The third stage of revision focuses on producing a final translation draft that is highly readable and usable. The content of this draft is accurate, and its terminology is consistent. All information has been included in the translation, and the appropriate house style has been adhered to. At this stage, revision is sometimes carried out by the same translator who did the initial translation. Alternatively, it can be done by a third party, thus allowing a fresh pair of eyes to spot any problems. This third-party reader can be a translator, but it might be an expert in the specialized domain of the source text *and* (crucially) a native speaker of the target language.

The new standard ISO 17100:2015 "Translation Services—Requirements for Translation Services" affecting such review processes has replaced the European EN 15038:2006 standard for translation service providers (TSPs). It includes a final check, labeled *review*, that is *intra*lingual (i.e., performed only within the same language) and focuses only on the specialized content of the target text. This review is optional. Moreover, the term *review* is intended to be distinct from the term *revision*, which refers to an *inter*lingual (i.e., performed between the source-language and the target-language texts) language check.

In contrast with a review, revision is compulsory and is always carried out by a second professional translator/editor. Based on external parameters,[2] different types of revision are possible according to the following:

- Degree of completeness, ranging from full reading to partial check and including scanning, spot-checking, and glance
- Types of error to be checked for: problems of meaning transfer, content, language and style, physical presentation, etc.
- Procedure involved, be it a predetermined sequence of steps for checking a translation, as adopted by the reviser: separate unilingual and comparative

checks; separate checks for errors relating to transfer, content, language/style and presentation; sequence of the checks, etc.

- Stage of the translation process at which the revision is carried out: either before the final delivery of the translation to the client—where it is a correction of errors with a pedagogical purpose or a quality control, in case it is carried out only on a portion of the translated text—or after the final delivery—where it is an identification of errors, known as *quality assessment* or *quality assurance* (Mossop 2014; Scarpa 2008, 221–222)

The process of translation in general, and revision in particular, can be long and time-consuming (see Figure 1.1). To minimize the time and cost of revision, the source text should contain as few errors in content, language/style, and/or presentation as possible.

The next sections discuss problems translators look for in the first stage of the translation process. They also review strategies translators often use to address them.

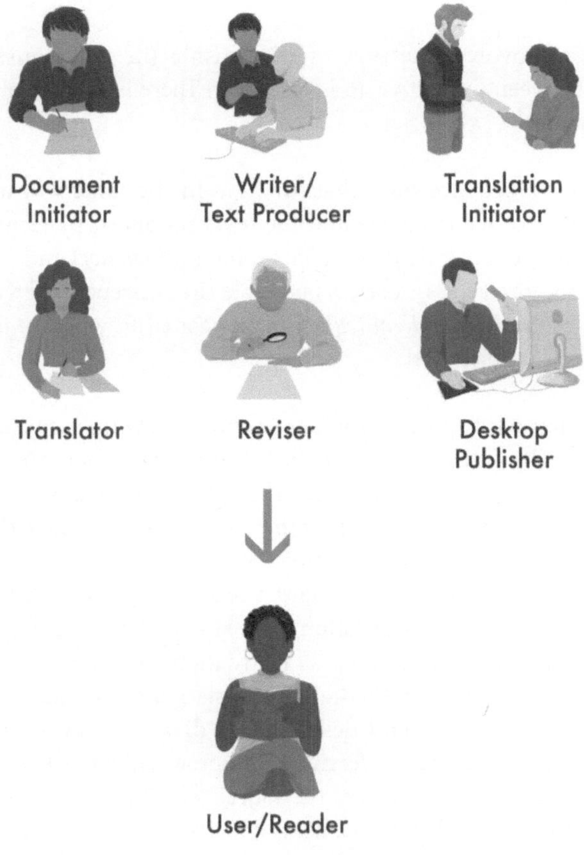

FIGURE 1.1 Professional roles in the translation process

Translating Technical Texts: Problems and Strategies

To adequately express meaning across languages, translators must

- understand the content of the source text,
- extract this information from the text, and
- express this information in the target text by using the words of the target language most appropriate to the new target context.

All of these have to be done despite differences in linguistic structure (grammar, terminology, etc.) and cultural gaps between source and target language. These gaps are inherent also in the translation of technical texts (see Kastberg 2007), though admittedly less so than in other translation areas. In technical translation the biggest issues are often about choosing the right words and register to suit the final users' needs of accuracy, readability, and usability.

Types of Equivalence

The main issue involves the type of relationship (i.e., the equivalence) that should hold between the source and target text. There is a constant tension created by balancing

- the ability to produce texts that function in the same way as the source texts in the source language and the texts produced by technical writers working in the target language (*functional equivalence*) and
- the rule to preserve as closely as possible the concepts (*referential equivalence*), wording (*formal equivalence*), and tone (*pragmatic equivalence*) of the source-language text.

The range of methods from a translator can choose lies on a continuum between two opposites. The first is *close* (or *literal*) translation, where the concepts are translated and the wording is retained. The other is *free* translation, where the concepts are translated without retaining the wording because the translator decides that the form of words resulting from a literal translation is not going to be sufficiently transparent for the target reader. (For more on this topic, see O'Brien, this volume.) Free translation gives rise to linguistic changes between source text and target text, known as translation *shifts* (Hatim and Munday 2004, 26–33). For example, in the Italian translation of the clause *This will end your Windows session* with "Fine della sessione di lavoro" [End of the work session], functional equivalence (referential and pragmatic) has been achieved by adequately conveying the content of the source text. However, this comes at the expense of formal equivalence, i.e., using the same wording in the source and target languages.

Equivalence can be achieved at the level of individual words (word-level equivalence) or clauses or even whole texts (above word-level equivalence). Exactly which one depends on the particular stretch of source text (the *unit of translation*) on which the translator focuses attention. To identify a translation problem and decide on the best strategy to solve it, the translator must often translate the segment as a whole in the target text (Malmkjær 1998, 286). In technical translation, the most typical problems are conceptual and lexical problems. In other words, the source language may express a concept that is totally unknown to the translator or that is not expressed by a corresponding term in the target language. Often, however, problems such as these can also be traced back to a lack of quality in the source text.

Equivalence at the Word Level

An example of a conceptual non-equivalence at the word level is provided by the translation in Italian of the job title *engineer*. The corresponding title, *ingegnere*, is awarded only after completion of a university master's degree, whereas in English *engineer* has a broader meaning. Conceptual problems at the word level can also be of a grammatical nature. That's what we see when the translator is not familiar with the typically compact syntactic patterns of technical terminology. For example, *mobile interface circuit* might be translated as "interface of the mobile circuit" instead of "circuit of the mobile interface." Similar problems can occur when the source text is badly written or is itself ungrammatical.

Other word-level grammatical problems can be posed by number, gender, and person (see Baker 2011, 105). For example, in Romance languages, adjectives have both number and gender, and the person system has a formality/politeness dimension that English no longer has. In English second-person, *you* is used to refer to everyone regardless of social standing. Both French and Italian, however, use the second-person singular *tu* to express familiarity, and French uses *vous* (second-person plural) while Italian uses *lei* (third-person singular) for an individual addressee in order to express social deference and/or non-familiarity. In Italian technical documentation, even more deferent impersonal or passive forms are preferred to address the reader/user. This is why, in the following example from a reference guide, all the *you* forms of the original have shifted to more formal passive constructions in the Italian translation (Scarpa 1990, 39; emphasis added):

> When *you* run the Setup program to prepare the application for use with *your* system, *you* specify what page size *you* normally use.
>
> IT: Quando *si* esegue il programma Setup per usare l'applicazione con il *proprio* sistema, *si* specificano le dimensioni di pagina *usate* normalmente.
> [When the Setup program *is being* run to prepare the application for use with *one's own* system, the page size *that is* normally *used* should *be specified*].

If we turn to lexical problems at the word level, we see that they typically occur with terms that are not used in everyday, ordinary language. Thus, a novice translator not familiar with the specialized field of the source text often does not know the field's terms either. These can be proper technical terms, such as *electrolytic capacitors*, but also, and more insidiously, terms that in the source language (but not in the target language!) have a specialized technical meaning different from their ordinary meanings. For example, the English term *fastener* has the same form but different meanings when used in structural steel design vs. everyday language. Italian, in contrast, has two different forms to signal the corresponding technical vs. non-technical meanings: *cerniera di fissaggio* in structural steel design vs. *chiusura* in everyday language.

Lexical problems can also arise when technical terms are used inconsistently in the source text. Such situations can occur when different terms are used to refer to the same thing or concept. An example of these issues can be seen in the use of terms such as a "drill-bit" in one section of a document but a "bore-bit" in other (for more on this topic, see Musacchio, this volume).

An example of a word-level lexical translation problem due to a lack of internationalization of the source text occurs in the translation of the names of the three separate help features *Search Companion*, *Wizard*, and *Assistant* in Microsoft Windows (Corbolante 2010). The challenge facing translators involved

- finding in their respective languages three distinct corresponding terms, all referring to animate objects (to match the corresponding icons appearing on the screen);
- avoiding a literal translation of "Wizard," either because in their languages it did not have the figurative meaning of "person with exceptional ability in a particular field or activity," which the word has in English, or because they deemed that it belonged to a register that was inappropriate to a technical context (see Table 1.1).

TABLE 1.1 The translation in French, Italian, Spanish, Dutch, German, and Swedish of *Search Companion*, *Wizard* and *Assistant* in Microsoft Windows. Source: Corbolante 2010.

	Search Companion	*Wizard*	*Assistant*
French	Assistant Recherche	Assistant	Compagnon
Italian	Ricerca guidata	Procedura guidata	Assistente
Spanish	Asistente para búsqueda	Asistente	Ayudante
Dutch	Zoekassistent	Wizard	Assistent
German	Such-Assistent	Assistent	Assistent
Swedish	Sökassistenten	Guide	Assistenten

Equivalence Above the Word Level

A lack of internationalization of the source document is also the cause of culture-specific problems above the level of the word in technical translation. This situation can involve idioms and fixed expressions, references to specific information and world-knowledge that are typical of the source language, a colloquial register, and a redundant and sometimes patronizing style. Cultural problems can be found in the more informal passages of a technical document, such as its introductory, parenthetical or concluding remarks. In many languages, including Italian, the standard strategies to translate idioms and fixed expressions in technical texts are either to render only their literal meaning or to make the colloquial register more formal and omit the redundancies in the style.

An example of these strategies is the translation of the subtitle in a reference guide titled *Want to Know More?* In Italian, it was translated as "Per saperne di più" and in French as "Informations complementaires." In both cases the colloquial register of the original, created by casting the subtitle as a direct question, has been toned down in the target texts by transforming the question into either a prepositional phrase (in Italian) or a noun phrase (in French). A further example is the following paragraph introducing a procedure where both the redundancies and the patronizing tone have been eliminated in the Italian translation.

> The following procedure describes one type of title page. Study carefully the use of the commands and how they work together in order to adapt this procedure to other types of title pages.
>
> IT Per ottenere qualsiasi tipo di frontespizio occorre eseguire la seguente procedura [To obtain any type of title page, the following procedure should be followed].

In this case, the translation ends up being more to the point and, atypically, more concise than the source text.

Quality in Technical Translation

Quality in translation is a relative concept, because it must be assessed on the basis of at least two competing criteria. The first hinges on the *accuracy* of the translation, which should express as exactly as possible the meaning of every word and turn of phrase in the original. The second is the parameter of *naturalness/readability* of the final result, which should not read like a translation at all but be as fluent and idiomatic as a native text. Accuracy is a source-text oriented parameter, because it is based on the idea that there is a link or bond of some sort between the source text and the target text (in other words, their *degree of equivalence*, on which the quality of a translation can be evaluated). On the other hand, naturalness/readability is a target-text oriented parameter, because the quality of a translation is evaluated mainly on how much the target text functions as an independent text in the target culture. The tension between the two parameters of accuracy and readability is of crucial importance in technical translation. In fact, in no other area of translation is the target text so unashamedly aimed at the target language readers and, at the same time, so closely related to the source text.

The Parameters of Quality

In technical translation, quality is achieved when the translation functions in precisely the same way as any other text in the target language. In other words, the translation sounds as if it was written by a native speaker of the target language. Technical texts are intended to function primarily for their target audience. Users of the target text are concerned about "getting the information they need and being able to understand and use it effectively in order to do something else, usually some task relating to their day to day work" (Byrne 2006, 15). By the same token, bad translation ranges between two extremes. At one end of the continuum, a word-for-word translation may be technically accurate, but the word order or choice of vocabulary is unduly influenced by the original language such that the sentences do not flow as smoothly as they might. At the other end of the continuum, a translation may be difficult to understand and often does not make sense, such that the user feels the need to refer to the original text.

Besides the textual parameters of accuracy and readability, translation quality in the language industry is also measured via parameters outside the text. These external parameters are more relevant to technical translation than to other areas of translation. The first of these is *adequacy*: the extent to which a translated text conforms to the set of specifications given by a client to the translator in the translation brief. The second is *acceptability*: the adherence of the target text to the norms and conventions of text production characterizing the text type of the translation in the target language. Especially crucial is the third, *usability*: the extent to which a translated text is efficient and easy to use.

In all cases, the quality of a translation varies according to its purpose, its urgency, and how much is to be paid for it (Samuelsson-Brown 2010, 105; Scarpa 2008, 214–215). For example, not all translation tasks require complete translations—or the highest level of quality. The client may be interested in getting only a translation of the most informative portions of the original document (selective translation) or a translated abstract containing all the relevant information (synoptic translation) or even only a quick oral summary of a written document.

The Translation Workflow

In the language industry, where translation projects can count in the millions of words, the translation process refers to the production chain of translated documents, the so-called *translation workflow*. Throughout the process, a project manager plans and supervises the various steps of the project and acts as a link among the various participants to the project. Other key stakeholders in the production of multilingual documentation are the following:

- The document initiator: the person or entity responsible for the production of the original source-language document (the author of the source text)
- The writer/text producer: an in-house or freelance contractor who produces the documentation on the basis of a brief from the document initiator
- The translation initiator: the person or entity responsible for starting the translation process by sending the document and some form of instruction (the translation brief) to the translator
- The translator: the person (an in-house staffer or a freelancer) or entity (translation/language service provider, localization vendor) responsible for the production of the target-language version of the original document
- The reviser: the person doing the revision of the translation (also called "reviewer," "proofreader," or "copy editor")
- The desktop publisher: the person checking the consistency of the layout of the translation with that of the original document;
- The user/reader: the person who is the real reason for the production of the original text (source-language user) and its translation (target-language user) (see Byrne 2006, 12–18)

In the translation workflow, the number of steps followed varies according to a number of variables, including the size of the TSP and the importance of the project. For example, the Quality Assurance Process shown in Table 1.2 was developed by a big language service provider and consists of ten steps, of which as many as five deal with revising the translation.

The same process-related considerations underlie the notion of quality management in international standards both general, such as the set of ISO 9000 standards, and specific to translation, such as the ASTM F2575–06 "Standard Guide for Quality Assurance in Translation" and the new standard ISO 17100:2015 "Translation Services: Requirements for Translation Services." These standards are not as much about the quality of the product/service proper as they are about quality assurance. (In other words, they focus on the set of procedures applied throughout the process to ensure that clients' quality requirements are met.) In the specific case of translation, linguistic quality is

TABLE 1.2 Ten-Step Quality Assurance Process (content adapted from www.ctslanguagelink.com/translation_process.php).

1. **Project Review**: review of the overall scope of the project: timeline, inventory of files, glossaries (if applicable), and final file delivery; identification of any typos, inconsistencies, missing text, etc. in the source document (or any other area requiring client clarification).
2. **Template Building/Translation Preparation**: creation of a template for the translation phase (if applicable), e.g. if the client's source file needs to be cleaned up or recreated if it is in a non-editable format.
3. **Glossary Development:** development of a standard glossary as a reference for the translators to be validated by the client.
4. **Translation:** ideally carried out by appropriately qualified and experienced translators who are native speakers of the target language and expert in the subject matter of the source document.
5. **Copy Editing:** review of the initial translation by a second professional translator/editor to ensure the quality of the translation as far as errors in grammar, spelling, punctuation, terminology and style.
6. **Translator/Client Review:** subsequent review by the original translator, who will accept or reject the suggested changes, and (if requested) further review of the translation by the client.
7. **Formatting:** performed by the desktop publisher to match the layout of the original material.
8. **Quality Control/Proofreading:** mechanical proof: layout and pagination, images, fonts, headers/footers, spelling of proper names, incorporation of source-language updates.
9. **Final Translator Review:** final review by a translator on the formatting of the document to verify hyphenated words etc. and check for any omissions from the text that may have been introduced during the final changes and formatting stage.
10. **Final Check/Delivery:** final verification by the Project Manager that the layout, page numbers, proper names and other details match the source document one last time.

therefore only one such requirement, together with quality of service and quality of presentation (Palumbo 2009, 99).

Given the connection between the quality of the source text and the quality of the resulting translation, the quality of translations as products rests firmly in the hands of both the translator/reviser *and* the source-text writer—hence the need for the translator to be involved in the creation of content from the start of creating the original source-language document. This type of collaboration between technical communicator and translator has been in place for some time in the localization industry, where translation does not follow the development of the product but proceeds side by side with it (Pym 2004, 16).

Training and Accreditation of Translators

All translators should be qualified and, ideally, hold a university degree in translation studies. Sadly, the reality of the translation market is that the job is often very poorly paid. Moreover, in most countries, anybody can claim to be a translator because there is no compulsory academic or professional qualification required to enter the profession (with the possible exception of sworn translators, approved by courts of justice). Becoming a translator, however, involves a complex mix of theory and practice or, to use Douglas Robinson's words (2003, 100–108), various types of experience, made up of intuitions (*abduction*), rules, models, laws, and theories (*deduction*), and the capacity to build patterns (*induction*). In order to get this composite experience, Samuelsson-Brown (2010) identifies the four following desirable requirements:

1. Attaining an undergraduate degree in modern languages or linguistics and completing a graduate degree or certificate in translation studies
2. Spending a period of time in the country where the language of choice is spoken and gaining a working knowledge of, or experience in, the specialized domains of choice
3. Getting a job as a trainee or junior translator with a company
4. Mastering the necessary IT skills and software used in translation (8–9)

Additionally, a detailed list of abilities and skills that translators should have outcomes when translators' employers are given the chance to have a say.

Within the European project OPTIMALE ("Optimising Professional Translator Training in a Multilingual Europe"), a Europe-wide survey was conducted in 2011 among translators' employers. Individuals were asked which competencies they deemed most important when recruiting new translators. The respondents were mostly commercial-sector translation service providers (TSPs) and translation departments or services of private or public companies. The results of the survey[3] stress the importance that employers give to both experience in the field of professional translation and a university degree in translation or related fields, with a slight priority being given to the first. What employers most desired in TSPs was the ability to produce 100% quality. Other important expectations were domain specialization, terminology extraction and management, use of standard tools and technologies, project management (e.g., define and/or apply quality-control procedures, define resources required, produce estimates), and client-relations competencies (e.g., identify client requirements, consolidate client relationships).

The premise on which the OPTIMALE project was founded is that translator training is the shared responsibility of universities and employers. This is basically the same approach as that subsumed by the European Master's in

Translation (EMT) network (http://ec.europa.eu/dgs/translation /programs/ emt/ index_ en.htm), a partnership project between the European Commission and, at the time of this writing, 64 higher-education institutions offering master's level translation programs. The EMT project, moreover, has established a quality label for university training programs that meet agreed-upon professional standards and market demands. Ultimately, it aims to enhance the status of the translation profession and improve the quality of translator training across the European Union.

The need to raise the quality of translation graduates can be traced back to the convergence of vocational and academic criteria in translator training that occurred in numerous countries in the 1980s and early 1990s. This situation resulted from a series of reforms that integrated technical or professional training institutions into national university systems (Pym 2001, 282–285; Kelly and Martin 2009, 294–297). This institutionalization of translator training involved the creation of an international network of university-level training institutions. Named the Conférence Internationale des Instituts Universitaires de Traducteurs et Interprètes (CIUTI), the network was established officially in 1964 with the aim of ensuring the quality of graduates from its member institutions.

While giving these traditionally vocational programs a greater legitimacy within academia, institutionalization has led to greater contact with other cognate non-vocational disciplines such as linguistics, comparative literature, and cultural studies. In Europe, this interaction resulted in the growth of university-level programs for translators. Such programs, however, focused less on translation's professional aspects. This situation led to the need for the European Union to try to harmonize, through the EMT network, the model for designing university translation studies programs focusing on the job's professional aspects. To create the model, participants developed a translator-competence profile listing all the competencies that translators need to work successfully in today's market (http://ec.europa.eu/dgs/translation/programs/ emt/key_documents/ emt_competences_translators_en.pdf).

Translators are also offered non-institutional training by professional associations of translators and major employers in the private and public sectors. Professional associations can, for example, organize short courses designed to address very specific aspects of translators' activity (new technologies, quality and revision, marketing, etc.) or longer-term programs. Such is the case for the American Translators Association (ATA) and the British Institute for Translators and Interpreters (ITI), both of which provide continuing professional development for their members to maintain their certification credentials.

In many countries, professional organizations also carry out certification and accreditation examinations and/or participate in university programs. In Australia, for example, the Australian National Accreditation Authority for Translators and Interpreters (NAATI) accreditation is the only credential officially accepted by employers for the profession of translation and interpreting.

Other short courses on very specific professional aspects of translators' activities are run by major employers in the private sector, such as larger translation companies, which often have agreements with universities offering internships to integrate theoretical learning with more practical skills.

In Europe, the need for continuing professional development and a trans-European certification for translators to ensure comparability, promote competitiveness, and increase employability has been relatively recently addressed by the European project TransCert (http://transcert.eu/). The innovative approach of the project involved all stakeholders in the translation sector (language industry, training institutions, and professional associations) in order to shape the translation business by consensus. The ultimate objective of TransCert has been to establish a voluntary, trans-European certification for translators to enable them to gain professional visibility.

Conclusion

Translation has become an important issue for businesses that want to market and sell their products in international markets. It has become part and parcel of their products. Within this context, this chapter has attempted to provide an overview of technical translation, which has the ultimate goal of producing a target-language version of the source text guaranteed to be relevant and useful for technical experts in the target culture. The process of translation has been discussed primarily in terms of a translator's thoughts and internal cognitive processes during the translation activity.

In addressing such topics, the chapter has used examples typical of technical translation to reveal how translators perceive translation problems and how they apply strategies to solve them. The chapter has also discussed the level of a translated text's quality and how it can be evaluated against various parameters both internal and external to the text. The process of translation has been analyzed within the language and translation industry, where it is seen as a multi-step translation workflow. Furthermore, the role of the translator in relation to that of the source-text writer has been discussed briefly to illustrate how the two must work in tandem to produce quality translations. Finally, the chapter has described the composite mix of competencies and skills that both institutional and non-institutional training should provide translators for their work.

Throughout this chapter runs the theme that the quality of a technical translation is closely linked to two interrelated factors:

1. The translator must have the ability to produce a target text that is an accurate rendition of the content of the source text and is as readable and easy to use as a non-translated text.
2. The quality of the source text is paramount for assuring the quality of the resulting translated text.

The latter requires a full collaboration between the source-text writer and translator, beginning in the very first stages of the production of the documentation to be translated. Corollary ideas resulting from the main thesis are the following:

- Unlike other types of translation such as literary translation, technical translation is always possible provided that
  - the translator has at least some degree of technical expertise in the particular subject matter of the source text and knowledge of its written conventions and/or
  - the source text has been written in a fully comprehensible way.
- As a rule of thumb, the clearer the information in the source text and the more transparent its style, the easier the translator's decision-making process and the speedier the transfer stage.
- Greater collaboration between translator and source-text writer in co-authoring the source text calls for a greater integration of the training programs and work practices of translators and technical/professional communicators (see Gnecchi et al. 2008, 2011).

In sum, the central idea to remember is that if good translation adds value to a product, a flawed translation provides disadvantages for both businesses and clients, because it damages the corporate image and affects the client's legal liability.

Despite the advantages that internet adoption around the world has brought, language still remains a challenge for businesses to reach a larger audience. Though English is perceived as the lingua franca of business communication, the number of non-English speaking internet users still exceeds that of English-speaking users, and citizens of many countries, such as China and Brazil, still prefer to converse in their local tongue. As the former German chancellor Willy Brandt once remarked, "If I am selling to you, I speak your language. If I am buying, *dann müssen Sie Deutsch sprechen*."

Notes

1 Technical translation is not the same as localization, though they are two different intersecting activities that partially overlap. Translation is the linguistic and cultural process of converting text from one language to another to make it appropriate to the specific socio-cultural context where it will be used. Besides the linguistic and cultural adaptation of the texts and visuals, localization involves a series of non-linguistic technological activities encompassing the development process of a product (software engineering, testing, project management etc.) (Esselink 2000, 3–4). Thus, localization is only one, very specific, area of the translation enterprise (Pym 2003).

2 Such parameters include the translator's level of competence in the specific subject matter of the text, the importance of the translation, the level of competence of the

readers of the translation with its subject matter, and how widely the translation will be used.

3 The results of this survey can be found in the relevant OPTIMALE Progress Report (public part) on the project's website (www.translator-training.eu).

References

Baker, Mona. 2011. *In Other Words: A Coursebook on Translation*. 2nd ed. Abingdon, UK/New York: Routledge.

Bellos, David. 2011. *Is That a Fish in Your Ear? Translation and the Meaning of Everything*. London: Particular Books.

Byrne, Jody. 2006. *Technical Translation: Usability Strategies for Translating Technical Documentation*. Dordrecht: Springer.

Corbolante, Licia. 2010. "La terminologia nella localizzazione." Presented for Ph.D. students, Trieste, Italy, May 13.

Durban, Chris. 2011. "Translation—Getting It Right: A Guide to Buying Translation." *American Translators Association*. www.atanet.org/publications/getting_it_right_trans.php.

Esselink, Bert. 2000. *A Practical Guide to Localization*. rev. ed. Amsterdam/Philadelphia: John Benjamins.

Gnecchi, Marusca, Bruce Maylath, Birthe Mousten, Federica Scarpa, and Sonia Vandepitte. 2008. "Professional Communication and Translation in Convergence." In *Proceedings of the IEEE International Professional Communication Conference, 13–16 July 2008, Concordia University, Montréal, Québec*, edited by Brian Still, 191–205. New York: IEEE.

———. 2011. "Field Convergence between Technical Writers and Technical Translators: Consequences for Training Institutions." *IEEE-Transactions on Professional Communication* 54, no. 2: 168–184.

Halliday, Michael A. K. 2004. "On the Grammar of Scientific English." In *The Language of Science*, edited by Jonathan J. Webster, 181–198. London/New York: Continuum.

Hatim, Basil, and Jeremy Munday. 2004. *Translation: An Advanced Resource Book*. London/New York: Routledge.

Kastberg, Peter. 2007. "Cultural Issues Facing the Technical Translator." *The Journal of Specialised Translation*, no. 8: 104–109.

Kelly, Dorothy, and Anne Martin. 2009. "Training and Education." In *Routledge Encyclopedia of Translation Studies*, 2nd ed., edited by Mona Baker and Gabriela Saldanha, 294–300. London/New York: Routledge.

LTC (Language Technology Center). 2016. *Expectations and Concerns of the European Language Industry*. https://ec.europa.eu/info/language-industry-facts-and-figures-lind-web_en.

Malmkjær, Kristen. 1998. *Translation and Language Teaching: Language Teaching and Translation*. Manchester: St. Jerome Publishing.

Mossop, Brian. 2014. *Revising and Editing for Translators*. 3rd ed. London/New York: Routledge.

Olohan, Maeve. 2016. *Scientific and Technical Translation*. London/New York: Routledge.

Palumbo, Giuseppe. 2009. *Key Terms in Translation Studies*. London: Continuum.

Pym, Anthony. 2001. "Translator-Training Institutions." In *Routledge Encyclopaedia of Translation Studies*, 1st ed., edited by Mona Baker, 280–285. London/New York: Routledge.

———. 2003. "What Localization Models Can Learn from Translation Theory." *The LISA Newsletter Globalization Insider* 12 2/4, May. www.translationdirectory.com/article449.htm.

———. 2004. "Propositions on Cross-Cultural Communication and Translation." *Target* 16, no. 1: 1–28.

Robinson, Douglas. 2003. *Becoming a Translator: An Introduction to the Theory and Practice of Translation*. 2nd ed. London/New York: Routledge.

Samuelsson-Brown, Geoffrey. 2010. *A Practical Guide for Translators*. 5th ed. Bristol/Buffalo/Toronto: Multilingual Matters.

Scarpa, Federica. 1990. "Trasposizioni stilistiche nella traduzione in italiano dall'inglese di software e documentazione tecnica." In *Studi sulla traduzione tecnica e letteraria*, 35–48. Trieste: Lint.

———. 2002. "'Closer and Closer Apart'? Specialized Translation in a Cognitive Perspective." In *Translation Studies: Perspectives on an Emerging Discipline*, edited by Alessandra Riccardi, 133–149. Cambridge: Cambridge University Press.

———. 2008. *La traduzione specializzata: Un approccio didattico professionale*. 2nd ed. Milano: Hoepli.

Wolpert, Lewis. 2002. "Is Science Dangerous?" *Journal of Molecular Biology* 319, no. 4: 969–972.

2

PRAGMATIC TRANSLATION AND ASSESSMENT FOR TECHNICAL COMMUNICATORS

Countering Myths and Misconceptions

Patricia Minacori

Introduction

This chapter begins by outlining concepts related to translation, particularly addressing "pragmatic translation," or translation for communication purposes, in the context of technical communication and excluding literary translation. A central objective is to demonstrate that translators are experts, and their role in the overall documentation process is as important as that of engineers. Indeed, translators may be considered as translation engineers, for, like engineers, they typically have earned a master's degree and gained a set of specialized competencies in various domains. They adapt to many situations, they become project managers, but unlike engineers their asset is communication, and they are very good pedagogues, i.e., they know how to explain even complex matters in a simple way. Translation is a process that relates first and foremost to meaning, as opposed to words. In that regard, it is fundamentally a process focused on comprehension. That is, translators seek to find the best way to convey the meaning presented in a text as it transits from one language to another vs. simply replacing words in one language for words in another. As this focus on comprehension—or understanding the meaning that a text conveys—underlies all others, it is essential to understand what translators do when they address the idea of comprehension in their work. In sum, the comprehension phase is the foundation of any act of translation, thereby allowing subsequent phases to take place. This chapter aims to state general comments about translation, show that translators are experts, explain the main aspect of translation (i.e., comprehension and the role of discursive parameters), and explain translators' competencies, translation procedures, and assessment.

To examine this topic, the chapter outlines the different phases of comprehension in detail, emphasizing realities that run counter to widely held myths and misconceptions by highlighting certain key points: First, translators are experts because they put into action a network of knowledge based on, and inspired by, the document they are to translate. In so doing, they also take into consideration its parameters concerning discourse. Next, translators are experts because they possess skills and competencies acquired over the course of their training and, especially, their careers. Finally, evaluating a translation (i.e., the document resulting from the translation process) is also the task of an expert translator who understands the nuances of the languages involved. The key is for technical communicators and for project managers to understand such processes and where they occur in the overall document production process. They can then draw on this knowledge to provide effective and targeted feedback at important points in the overall process. The result of such informed interactions are reliable, high-quality translations.

General Comments about Translation

In writing about "translation," it is important to note that there are numerous concepts concerning how to do translations. Additionally, the term "translation" refers to both the translated text *and* the process by which one arrives at this text. If one considers the history of translation in the West (Ballard 2007), the evolution or processes can be understood in terms of the following paradigms:

- A word-for-word paradigm, seeking faithfulness to God's Word; this is embodied in early translations of the Bible (focus on words);
- A creation paradigm, in particular in Cicero's *De Oratore* treaty on translation, which aimed to replace word-for-word translations with use of the best expressions in Latin so as to respect the eloquence of the orators (focus on style);
- A meaning paradigm employed in Saint Hieronymus's new Latin version of the Bible (385–420 AD) in which meaning for meaning is chosen (focus on meaning);
- A text intention paradigm; in France, Etienne Dolet and his five translation rules (1540) opposed word-for-word translation and sought to express the author's intention (focus on intent);
- A creation paradigm encapsulated by the French so-called *Belles Infidèles* ("beautiful but unfaithful") movement—this Renaissance movement, initiated by Nicolas Perrot d'Ablancourt, advocated a turn away from literalism (focus on ideas);

- A paradigm of fidelity to the source text during the Romantic period (Ballard 1999) (focus on the source text);
- A decision paradigm attributed to Levý (1967), which proposed to discuss pragmatic translation (focus on pragmatism).

These paradigms reveal that translation has historically moved in one of three directions.

In the first, the author is the focus for the process. In the second, the recipient (i.e., the person who will use information in the translation) is at the forefront. In the third, translation becomes a creative work, highlighting the translator. In this third approach, the translator focuses on creation, instead of concentrating on the intentions of the author or the constraints of the recipient and she/he completely disconnects from the first two directions.

According to some experts, translation is also a moving object. For example, Boulanger (2004, 58) observes:

> Over the last thirty years translation has come to be emancipated—it has gained in self-confidence and little by little it has left behind the scientific sphere of linguistics, opening the way to interdisciplinarity with other humanities.
>
> *(author's translation)*

The central point Boulanger seeks to make—and that is central to this chapter—is that translation studies is a central topic of research.

In the professional world, there are multiple producers of texts, which means that there is a variation in the priorities which may be emphasized in the text being translated. For example, a publishing house that publishes literary or editorial translations holds quite different priorities from a firm interested in technical documentation. The same holds true of international organizations, governments, trade unions, and so on. Even within an organization or firm, translations also span a wide range of different subject area, including economics, advertising, law, science, technology, politics, diplomacy, and film subtitling. Owing to this plurality of objects and multiplicity of objectives, one can see that it is difficult to approach all of these translation practices in a homogeneous way.

If the act of translation represents a wish to communicate, then relying on a single translation strategy will not produce an altogether usable text. In the case of a literary text, the aesthetic will take special priority, and it will often be necessary for the translator to distinguish between the meaning's effect (i.e., how the author wishes to influence the recipient's feelings and emotions) and the meaning itself (i.e., what the words actually say). In the case of a UN resolution, the translator's priority will be to take all necessary diplomatic precautions in expressing the statements found in a resolution. And in the case of instructions, the translator will need to aim for clarity, precision, and terminological coherence so that the text is "user friendly" and leads the reader to complete a process safely and efficiently.

With these factors in mind, it is still possible to discern several aims which underlie every translation; these include

- to inform, as in the case of economic or financial results, or of scientific and technical articles;
- to convince, as in promotional or advertising-type documents;
- to entertain, as in video games;
- to specify, as in users' manuals and instructions, help guides, maintenance guides, and patent norms;
- to teach, as in student manuals and e-learning modules;
- to suggest a performative act, as in accepting at a mouse click the general conditions of sale for a product or service on the internet.

The categories here do not represent rigid distinctions. Translating a firm's financial results can, for example, both inform and convince a potential investor to act (or not).

If one uses a spatial-temporal framework (i.e., a situation at a certain point in time), that individual can determine certain paradigms of translation, including

- literary translation, which seems to focus on the author of the original text and the person for whom the translation is destined as guided by a clear editorial policy; while communication is a central focus, aesthetics is also of central importance;
- pragmatic translation, which aims to communicate in the context of a globalized business environment, especially in order to persuade (consumers to buy), to specify (in user manuals), to teach (as in e-learning modules), to entertain (via video games); it may also have a performative objective (legal contexts);
- diplomatic or political translation, as can be seen in international organizations, where the translation involves taking precautions to use words carefully, thereby respecting difficult discussions leading to sometimes delicate, negotiated positions.

With this framework for translation established, the next step is to consider the degree of the translator's expertise. Technical communicators can understand a translator's expertise better than most because they share certain skills with translators, such as competence in writing for a specific end-user. However, other skills differ considerably. For the sake of a successful communication project, technical communicators need to know what these other skills are and how to assess them.

It is vital that technical communicators recognize that they are integrally involved in producing high-quality versions of the same documents within the same documentation supply chain. Thus, technical communicators and translators must participate in a constant conversation about the documentation on which they are working. To do so, they must have at least a rough idea of each other's skills. For example, technical communicators need to know about the process of translation. Translators, in turn, need to have some understanding of the composing and design of the documentation that they will translate and localize. To this end, the two groups need to have some understanding of the objectives and processes involved in each other's work.

Translators Are Experts

It is important for technical communicators to understand how the documentation they have produced will be translated. To help clients new to translation with the basics, la Société Française des Traducteurs and the American Translation Association have published, in French, English, and other languages, an easy-to-use pamphlet, *Translation: Getting It Right* (Durban 2011).

It is important for technical communicators to understand the falsehoods underlying common myths and misconceptions of translation so that they can work more effectively with translators based on a knowledge of what they actually do. A particularly common misconception is that anyone who knows two languages can translate. However, such a person cannot act as an impromptu translator. Rather, translation—and the translation profession—is founded on solid knowledge of tested methodologies and numerous competencies. Moreover, these are acquired over the course of educational degree programs and a career and not from simply speaking two or more languages. The failure to understand such distinctions can result in considerable and costly problems in documentation for an organization. The central point to remember is that translation is a skilled technical profession that requires an in-depth understanding of processes, practices, and topics. (A detailed description of the competencies required of professional translators appears in a later section entitled "The Translator's Competencies.") Thus, the idea that "anyone who speaks multiple languages can be a translator" is akin to "anyone who knows how to write can be a technical communicator."

Also problematic is the assumption that machine translation (MT)—or the use of certain computing software to do translation—can render meaning accurately. The free software market, available on the internet, offers widespread examples of tools purporting to replace a human translator. The myth that such translation tools can meet a legitimate business's translation needs reflects a focus on savings, rather than a grip on reality. The myth can be maintained only with a total lack of knowledge concerning the difficulty of translation and a total disregard of the recipients' needs when using the target (i.e., translated) text.

Moreover, translators often are brought in at the end of the document creation process and at a point just before product release or publication. The problem here is time. Effective translation cannot be done with the wave of a magic wand; rather, it takes time. The average translator, for example, can translate on average three hundred words an hour, and this factor affects how quickly translated materials can get to international audiences (and markets). A report published by le Syndicat National des Traducteurs Professionnel (Commission Statistiques et Étude du Marché 2010) bears this out in responses to a survey conducted in 2009. Of the 1,057 translators who answered questions about their jobs, in particular about their productivity, 35% said that they translate between 200 and 300 words per hour; 30%, between 300 and 400 words per hour. With roughly 300 words corresponding to the length of a text on a page, we can reasonably say that translating a page per hour is about average.

Certain firms tend to develop products made from reliable materials, create pieces of software responding to a real need in the market, and draw up documentation according to best practices so as to sell the product. In so doing, they reveal that they take the user seriously, and they thereby ensure that their product responds to users' needs. Firms that care about their users recognize that translation must be carried out by an expert possessing an array of competencies. Thus, the more technical communicators understand what translation is and what translators do, the better they can work with them to produce the kind of high-quality documentation that can benefit—and add value to—their organization.

To help with this objective, the following section overviews what translators do connected to comprehension: one of the most fundamental phases of the translation process. By understanding such factors, technical communicators will be able to work more effectively with translators. Technical communicators can also use this understanding to create materials that are easier for translators to translate quickly and effectively as they have been designed with an understanding of the translation process.

Comprehension in Translation

The myth is so persistent, it must be reiterated: *translating does not involve translating words*; rather, a translator strives to find an equivalent way to convey the meaning intended by the author of a text. To translate the expression, "It is raining cats and dogs," word for word renders a confusing sentence; translating that same text for meaning gets to the focus of the author' intentions: that the author means, "It is raining very forcefully." The objective is comprehension: Does the reader correctly comprehend the meaning of the ideas that the writer wishes to convey in the text?

When we speak of comprehension, we really are speaking of comprehension at different levels. Of all the comprehension models studied, the situation model created by Van Dijk and Kintsch (1983) is particularly interesting in relation to meaning. The central idea is that an adult reader's comprehension intersects on three levels:

1. The surface level: corresponding to the entry message, it is when the reader views the text in its lexical components, syntax, and so on;
2. The semantic level of content: This is a text's micro- and macrostructures; and
3. The situation model level: This includes general knowledge of the subject and its context and allows the inferences that the reader needs to make in order to understand the text.

In practice, these three levels work as follows:

Level 1: Surface Level

On the surface level, readers decode a linguistic form, which corresponds to an entry message. They cast their eye over a chain of letters, which allows them to decode the signs (e.g., seeing the letters "W-O-R-D" and recognizing they represent the term/idea "word"). For the decoding to take place as quickly as possible, it needs to be as automatic as possible. The other operations, which are understanding on a) the semantic level and b) of the situation model, can then take place.

Level 2: Semantic Level

On the semantic level, content is made up of microstructures corresponding to an entire set of propositions that represent the surface of a text. To use an example from Kintsch (1974), when a reader reads the words, "The dog is barking in his kennel," the microstructure is, "In [kennel, barking (dog)]." However, this microstructure does not exist in isolation. Rather, it is part of a higher level called a macrostructure, organized into paragraphs and sections. To construct meaning accurately, readers must interpret the microstructures within the context of their surrounding macrostructures. When readers attempt to summarize a text, they have to take into account the text's macrostructures; in other words, readers condense and organize the microstructures by putting them into a hierarchy.

Level 3: Situational Model Level

The situation model level allows readers to accede to meaning and to understand how a text coheres, with a beginning, middle, and end. A word or a group of words cannot be understood alone. Readers need to take into account the surrounding context: that is to say, the words which come before or after. Let us take as an example what has to happen when professional English-to-French translators read the following title: "Water Repellent Coating Could Make Power Plants Greener." (This is a real title of an article in *MIT Technology Review*: Becker 2014.) Because the translators are professionals, they know that a power plant is a facility producing energy. If a power plant is greener, it means that it is efficient. The term "Greener," coming after "Power Plant," is clear; professional translators know the different concepts exposed in the text. The context helps translators understand. But here's the rub: This is a text that I gave my undergraduates to translate, and to my great surprise (and dismay), four students out of 26 did not understand the context because some concepts were unknown to them. They thought that "plant" meant some kind of vegetation because of its lexical proximity to "greener." This misunderstanding led to a total incomprehension of the whole text and a complete absurdity when they translated it.

The situation model describes a person's cognitive representation of events, actions, and the situation evoked by the text. It takes several types of knowledge into account:

- Readers' general knowledge, derived from life experience and previous readings and discussions of topics covered in the text. In the preceding example, experienced technical translators would have known what a power plant is and the challenges of designing power plants for efficiency.
- Readers' knowledge of language and rhetorical patterns (including tropes), developed with practice over time; this helps readers understand what is implicit in the message and what inferences to construct—in other words, to think rationally so as to re-establish the logical links in the text. Reasoning by analogy is of special importance. Indeed, each time that readers establish a link between what they have read and what they know, they employ a process of reasoning by analogy. In the power plant example, experienced professional translators read the text and know instantaneously that the topic of the article has to do with a facility, not a vegetable.

Finally, the text's internal dynamic is of fundamental importance in the way that readers deal with the information it contains. Comprehension is a phenomenon in constant evolution: Each time that a reader moves forward in her or his reading, that reader is constantly integrating knowledge of what she or he has just read with what she or he already knows in order to understand the next part of the text (Spiro 1977).

Additionally, a reader has to process the information in a text in different ways at two levels:

- Low-level forms of processing are necessary for comprehending the meaning of a word or comprehending the grammatical and syntactic organization. In the example of the power plant, professional translators must be aware that a power plant is a facility.

- High-level forms of processing allow a person to integrate pieces of information to other information by using, for example, context to comprehend the meaning intended by an author, or references to previous knowledge. The reader thus makes links with what she or he already knows in order to better understand new information. In the power plant text, professional translators know that, in the present era, achieving energy efficiency is a paramount challenge. The article explains new research investigating a new type of coating on a part of the facility, the steam condenser. The author explains why the coating fosters efficiency. To render the meaning accurately in the target text, professional translators link this new information with what they already know to understand the new information and thereby make it understandable to the readers of the target text.

All of these activities are governed by rules of efficiency: In order for comprehension to occur, it is necessary for the task to be carried out as well as possible with the least possible cognitive cost.

The more automated a mental task, the less costly it will be in a cognitive sense. A reader's comprehension in his first language is complex, because in order for it to be successful, the reader must

- automate the low-level operations to achieve fluency, that is, to read quickly and fluidly;
- employ high-level operations, that is, not to just decode the words but to make immediate inferences and connections.

According to Coirier, Gaonac'h, and Passerault (1996), the situation of a reader in a foreign language is just as complex, and she or he must proceed in a similar fashion by operating at both low and high levels of comprehension. In addition, however, as a bilingual or multilingual reader, the translator brings a special understanding of language to the text. And among bi- or multilingual readers, translators are readers of a still more distinct type. Even if reading in a foreign language employs the same mental processes as reading in a first language, translators operate in a pivotal situation in the communication chain, swiveling between an expert reader in one language and an expert author in another language.

In fact, translators read in an extremely specialized manner. Although translators read the source text to be translated, this text is never destined for them. For this reason, translators use a set of procedures designed to tease out the meaning of the text and discern precisely what needs to be communicated to the reader in the target language. This set of procedures is learned over a long period of instruction: In Europe, the standard training of translators takes five years, culminating in a master's degree in translation.

Thus, professional translators always begin by weaving together a framework of knowledge based on the text to be translated, and in doing so, translators have to consider the discursive parameters ("discursive" being the adjective form of "discourse"). While it is not essential for technical communicators to understand all of these items in full detail, an initial familiarity with what they are and how they guide translation practices can be key to working more effectively with translators to produce final texts. Thus, we can view a text as a written discourse representing an author's point of view, aiming to achieve various objectives with specific recipients in mind and published in a given medium at a certain date. From a translation perspective, these elements are discursive parameters.

Discursive Parameters

Discursive parameters allow translators to assemble an entire store of knowledge based on the text to be translated. They also allow translators to take into consideration a range of pertinent information to guide thinking about the text. Of central importance is the aim of translating the meaning of the text and not merely the words themselves. These parameters are

- the author,
- the recipient,
- the objectives,
- the medium (paper, website, etc.), and
- the date of the text.

These items are discussed more fully in the following sections so that technical communicators can better understand what they are and how they relate to the translation process.

The Author of the Source Text

In the case of technical communication, a text's author might be a "subject-matter expert" (SME) or a technical communicator working with SMEs. When the author is an SME, she or he and the translator generally do not share the same knowledge. The difference in the first case is that the SME knows the measures to take for a product to function or the steps to take to change the design of a piece of software. The translator, however, needs to know how to use the product or software. The SME has a thorough knowledge of the product but often only a sketchy knowledge of the methodology necessary to produce documentation for it. Conversely, if the author of the documentation is already a technical communication specialist, that person has already collected, organized, engaged with, and processed the documentation. In such cases, the technical communicator has still deeper knowledge about the product than the translator does.

The translator's ability to communicate with the author can be crucial to the quality of the translation. If the translator is part of a firm's or institution's organization, she or he can quickly contact the SME or author to obtain more information if needed. If the author and the translator are not located in the same town, country, or time zone, the translator can still produce a more accurate translation if she or he can contact with the document's author, even if collaboration involves web platforms. In both cases, the translator can ask questions and propose improvements to the documentation. Indeed, this assumption flows from the results of a study concerning 23 firms on the subject of exchanges between technical communicators and translators (Minacori 2011). Its findings indicated that, when there is close collaboration between technical communicators and translators, the resulting documentation is much improved. To ensure such communications, some organizations have developed documents that prompt answers to the questions that authors and translators often have to ask. For example,

- they establish a list of technical terms to be used (e.g., "sensor" instead of "captor"),
- they address situations when synonyms exist for the same concept by establishing a single term that harmonizes the text and eliminates the possibility the reader might presume that two terms might have two separate referents,
- they identify terms that could lead to confusion once translated (due to lack of precision) via a list of preferences; for example, several French options can exist for the English term "when": these are *dès lors que, lorsque, dès que, si.* The first, *dès lors que*, may be considered too literary. Thus, the list would show a marked preference for *lorsque* or *dès que, si.* The objective of marking these preferences is to facilitate the task of the translator.

In this example, the translators and the technical communicators have different but highly complementary skills. If opportunities for feedback exist between the two, a type of synergy is created, resulting in documentation that is well thought out and constructed to be more efficient. However, if the translators are not an integral part of the company, it will be more difficult to ask questions and improve the documentation, and the feedback loop will be broken. It is therefore vital that those responsible for documentation know that communication between translator and author is central to facilitating translation and ensuring the quality of the resulting documentation.

To establish such relationships, managers of documentation projects can bring translators to the production site to meet the experts involved in the communication chain (Minacori 2011). This allows them to see the products, understand the important issues, and ask questions. Such meetings can create opportunities for feedback between the actors in the documentation chain to improve the translation service, thereby achieving savings of time and money. For this reason, project managers need to understand what translation entails, so that they can look for and create opportunities for such feedback. They also need to anticipate difficulties and implement good working conditions that strengthen links between technical communicators and translators.

The Recipient of the Translation

The second parameter is the *recipient* of the translation. Knowledge of the recipient allows translators to adapt their translations to the reader's expertise in relation to the topic. If, for example, the recipient is an engineer or a technician or the general public, technical communicators adapt the documentation to the recipient's knowledge of the topic. In general, the technical communicator will have formulated the text with an audience in mind. (User-experience specialists do this by creating personas of their audience members.) Yet, it is not unusual for the translators and the recipients of the translation to come as an afterthought in this formulation. This can be problematic, especially when the translators suggest modifications to simplify the wording or to offer explanations if they determine that the users of the translation might need them. Accordingly, parties can avoid potential difficulties, curtail costs, and prevent delays if translators contribute to the documentation chain early on. Doing so facilitates feedback between the author and the intended recipients allowing author and translator together to weave a network of knowledge critical to accurate translation.

Familiarity with the recipient also allows the translator to recognize if there is a difference in knowledge between author and audience and if research is needed to fill potential knowledge gaps. Professional translators possess sharp research skills. They know how to find texts to help solve difficulties in comprehending the source text or expressing a concept in the target text. Monolingual and bilingual dictionaries, aligned lexical corpora, previously translated technical documents, subject-matter books, articles, and internet sites are all tools that translators readily employ to ensure that translations are accurate and precise.

Finally, knowing the recipient allows translators to test choices that they have made with potential recipients via interviews to confirm translation options. In "pragmatic translation"—which aims to communicate in a globalized business environment—one often hears the argument that the recipient is a discursive parameter of the highest importance (i.e., you never write or translate without having in mind a given recipient). Accordingly, the information one can gather about the audience directly affects the quality of the translation.

The Documentation's Objectives and Medium

In general terms, the objectives of pragmatic texts are to a) inform, b) convince, c) entertain, d) specify, e) teach, and f) suggest a performative act. One can also link the objectives to the medium of a text and how the recipients engage with it. For user manuals, help guides, or maintenance manuals in paper or electronic form, the translator's role includes rendering the text in its most readable form and ensuring the consistent use of terminology.

For marketing materials, translators draw from their first language to generate precise and concise phrasing to convince potential recipients that they must buy a product. Translators use not only effective terminology but also rhetoric (i.e., stylistics and structure of presentation, adapted to the reader of a specific language and culture). One does not appeal to the French-speaking public in the same way as one does to its English-speaking counterparts. For example, if we compare the layout of the front pages of newspapers in the United States with counterparts in France, we see that the journalists, editors, and page designers in the two countries take different approaches. US newspapers appear to contain more information, while French newspapers have more pictures and less text. For a translator, adjusting to address such differences involves adapting a message rather than translating it. As such, the translator needs to understand the importance of the adaptation to effectively target the person at whom the product is aimed (Andréani et al., 2008, Quillard 1999). For websites, translation alone does not suffice. Rather, broader localization is needed. Adapting information to users of a specific language and culture makes the language translator also a cultural localizer. In truth, translators have always held this role and engaged in localization even if the myth is that they work only with words.

Translators must also consider and adapt to the legal situation in each country, including laws governing the use of the internet, and a country's consumer norms, such as those regarding currency, color codes, etc. These can be highly sensitive and crucial for effective localization. For example, websites that prominently feature green are associated with religion in the Arab world. In France, this color is often associated with environmental awareness and renewable energy. In Thailand, by contrast it is often preferable to avoid the color green because many Thais associate it with the jungle, and thus diseases.

These examples illustrate an important point. Translators bring important cultural and linguistic knowledge to the work they do. Such knowledge, moreover, is acquired throughout a lifetime of being part of a culture. In this regard, translators have knowledge and skills quite distinct from those of technical communicators. (For more on localization, see chapters by Valli, by Mousten, and by Mousten and Riordan in this volume).

The Date of a Translation

Cultures can also use different formats for conveying information about dates, and translators need to be aware of and address such factors in their work. For example, 9.11.2001 is generally interpreted by Americans as September 11, 2001, but French audiences could read it as 9 Novembre 2001. (That is, readers in each nation place the numeric representation for the month and the day in a date in a different order.) Translators must therefore know about local conventions for writing dates and adapt to them. A similar situation is true when source texts are written some months or years before the translation and mention dates. Once again, the translator has to adapt the target text according to the date.

The Translator's Competencies

European institutions have long mapped skills and competencies required of translators because, for the 28 countries (soon 27 following Brexit) of the EU, encompassing 24 languages, every document is translated into every language. The importance of translation is thus demonstrated daily in the thousands of pages produced so that the European Union can operate openly and effectively in the languages of the speakers of its constituent member states.

Within this framework, European institutions are ambassadors for multilingualism, which they define as follows:

Multilingualism—working definitions:

- An individual's ability to communicate in several different languages
- The co-existence of different language communities in one geographical/political area
- An organisation's policy choice to operate in more than one language (European Commission 2015)

The EU is also involved in supporting high-quality training in translation with a view to employing competent translators for its institutions. Therefore, much importance is given to the translation profession, for it allows for effective communication across the Union's member states. Indeed, since its creation, the EU has declared that each of its languages carries equal importance. For this reason, these authorities have established a quality label, the European Master's in Translation (EMT), awarded to European universities that satisfy specified criteria.

Within the framework of the EMT, 48 competencies have been identified, studied, and grouped into six categories, covering the vast spectrum of translators' professional activities today. The EU defines the term "competence" as follows:

> By "competence," we mean the combination of aptitudes, knowledge, behaviour and knowhow necessary to carry out a given task under given

> conditions. This combination is recognised and legitimised by a responsible authority (institution, expert).
>
> *(Gambier 2009)*

The six categories of competence, in turn, are:

- language competence,
- intercultural competence,
- information mining competence,
- technological competence,
- thematic competence, and
- translation service provision competence.

This list reveals that when a profession is recognized for its true value, then translation receives serious study, garnering high-quality training of practitioners, taught by experts with deep knowledge of what is at stake and of how communication across languages functions.

The research group PACTE [Procés d'Adquisició de la Competència Traductora i Avaluació] (2003) has focused on competencies by developing a model for a successful translation that incorporates a system of knowledge with specific characteristics. First, in contrast to the misconception any bilingual can translate, a translator's expertise differs from that of a bilingual person by being made up of five inter-linked sub-competencies summarized as follows:

- Bilingual sub-competence, noting the translator has pragmatic, sociolinguistic, textual, lexical, and grammatical knowledge in at least two languages—the source language (of the original text) and the target language (the language of the intended translation), which is generally the translator's first language;
- Extralinguistic sub-competence, according to which the translator has encyclopedic and thematic knowledge in two cultures;
- Process sub-competence, which allows the translator to navigate the processes, methods, and procedures at the heart of translation work;
- Instrumental sub-competence, with which the translators can tap a wide range of documentary sources and computerized resources specific to the translation profession;
- Strategic sub-competence, most important of all, because it allows the translator to understand any difficulties encountered and how to resolve them by always aiming to use methods with the highest level of efficiency; this sub-competence allows the translator to plan and carry out the translation as well as evaluate the methods employed.

Second, the PACTE Group has added psychophysiological components as a consideration. They relate to all cognitive and behavioral mechanisms used in translation—memory, attention, perseverance, the ability to think critically—as well as psychomotor mechanisms. Notable here is the systematic character of these sub-competencies, which are the fruit of an apprenticeship and of long practice.

This systematic character reveals these competencies to be inseparable from the other. If one of them is missing, there cannot be a successful translation. Especially interesting about this model is that it defines competencies by virtue of the procedures that underlie translation. Translators' skills and competencies can be compared favorably with those of technical communicators (Minacori and Veisblat 2010). At this point we can turn to our attention to the different elements involved in the process of translation, which arise from the previously mentioned competencies.

Translation Procedures

Translation succeeds only when the *translator* comprehends a source text's meaning (not merely decoding the individual words) and renders the target text such that readers of the target language comprehend as much of the meaning as can realistically be conveyed. Only after the translator has achieved comprehension will she or he then be able to express this comprehension in the target language in a way that is fully consistent with the meaning intended by the author—all while keeping in mind the parameters of precision, rigor, and conciseness.

In terms of these ideas, Federica Scarpa's earlier chapter identifies and describes the persons involved in translation procedures designed for quality control. In this section, I delve further into these persons' specific activities. My goal is to show technical communicators how such factors are essential to avoid the errors that result when translation clients believe the myth that a usable translation can be rendered by anyone who has grown up with two languages or it can be accomplished by a secretary flipping through a bilingual dictionary, rather than hiring a trained, professional translator. While the translation process is complex (especially for a monolingual client), one can conceptualize—and needs to remember—the following steps (as depicted in Figure 2.1):

1. The translator reads the source text in the source language.
2. The translator gains an understanding of the logic, different parts, objectives, and other elements, expressed in a lexical structure (i.e., a set of words); the translator grasps the source text's grammatical rules, concepts, metaphors, and so on.
3. The translator begins to realize that she or he does not fully understand some parts or elements; a knowledge gap becomes apparent, separating the text's author from the text's translator. The translator becomes aware that she or he must carry out research to fill the gap.

FIGURE 2.1 The stages in the translator's translation process

4. The translator conducts research pertinent to the text to fill the gap; this is done by first investigating the text's context and then by researching other texts (hyper-textual research)—that is, by consulting monolingual or bilingual dictionaries but also relevant books, articles, internet sites, technical documents, etc.
5. The translator completes the research and reaches complete comprehension of the text.
6. The translator expresses her or his comprehension while respecting the author's intended meaning and the text's objectives with precision and conciseness and according to the reader's expectations. (This last concern applies to technical communication but differs greatly from literary translation, in which the author's presence can be magnified.) In this step of the process, supplementary research might be necessary.
7. The translator must go back and revise the initial translation. The word "revise" is employed here rather than "proofread" because revising is truly seeing again in a new way. The distinction can be seen in the definitions of the two verbs: "Revise: to examine or re-examine and improve or amend" (*Concise Oxford* 2005, 1180) "Proof-read: read and mark any errors" (*Concise Oxford* 2005, 1097).

Improving the text through revision, moreover, goes well beyond reading and marking errors. After the initial translator revises the initial translation, typically a second professional is brought in to review the initial translator's target text. This translator, known in the profession as a "reviser," is *de rigueur* for any organization interested in translation quality control.

Analyzing EN 15038, a European quality standard for translation service provision, Palumbo (2016) lists professional competencies for revisers. As second translators, their tasks involve

- considering source and target languages;
- looking at terminology, register, and style; and
- recommending corrective measures.

Following revisers come reviewers. As they typically lack translation experience, they are considered domain specialists, not translators. They check the target text to make sure that it respects the conventions of the domain.

Palumbo (2016) also mentions the ISO 17100:2015(en) Standard for Translation Services and Requirements for Translation Services, in which "the term bilingual editing is sometimes used as synonym for revision." Such revision can be divided into three phases (as depicted in Figure 2.2).

Phase One

The reviser compares the target text to the source text to see if the translation expresses each message intended by the author. In other words, the reviser checks to see if every meaning in the source text is present in the target text. In this phase, the reviser checks to see if any phrase, sentence, or even a full paragraph has been omitted, as well as to see if the author's intention has been fully respected.

Phase Two

The reviser focuses on the quality of the target text. Perhaps nothing has been omitted, but the ideas in the text can be expressed in ways that are more easily comprehended in the target language. On reaching this phase, the reviser needs to put aside the source text to remove its linguistic outer shell. This allows the reviser to concentrate solely on the target text's quality. In this way, the reviser

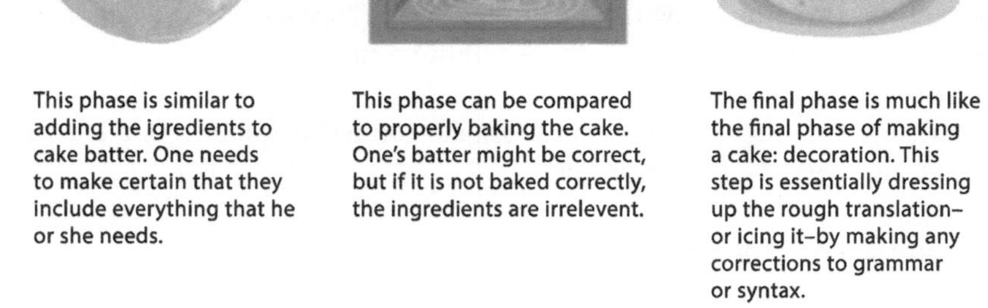

FIGURE 2.2 The phases of the translation reviser's work

avoids being influenced by the source text's grammatical structures and lexical cognates, which can seep into a translation and make it sound unidiomatic to readers of the target language.

During this phase, revisers must beware of calquing: the process of simply translating a phrase word for word, thus forming a "calque" (as someone relying solely on a bilingual dictionary might do). Relying on a quick calque works against the goal of recognizing that the target language expresses the concept of the phrase in different and more idiomatic phrases. For example, someone without professional training might be tempted to translate the French phrase *légitime defense* as "legitimate defense." In actuality, the phrase needs to be translated as "self-defense." In addition, during this phase, the reviser checks the cohesion of the text as well as references and terminology to see if they best match standard practice.

Phase Three

For this final phase, the reviser focuses on making improvements to grammar and syntax. Translators are taught to separate Phases Two and Three so that their working memory is not impaired when checking form vs. substance.

In this three-phase approach, revision adds the icing on the cake, but it also makes sure that the cake is fully baked. To ensure that the icing too is of top-notch quality, many organizations add a "taste tester," so to speak. Often called a *reviewer*, that person is usually an intended reader or someone very close to the intended readers, such as a member of the client organization's sales or marketing staff. As mentioned earlier, frequently the reviewer does not know the language of the source text at all. However, the reviewer can flag words and phrases that sound odd or unfamiliar to a native speaker of the target language.

With this dissection of the translation process, we have seen the complex nature of translation, particularly a technical translation. It is not merely the product of a secretary wielding a dictionary or even of a person who happens to know two languages. Rather, it is the work of professionals, often numbering at least two, with the second checking the work of the first. A peculiarity of translation is that the client who commissions the translation rarely can evaluate what he or she has bought. The only way to evaluate the delivered product is to hire yet another translator to judge the solidity and quality of the translated document. In fact, a person who is not a professional translator of the languages involved simply does not have the mindset or competencies of a translator to evaluate a translation accurately.

Translation Assessment

This section is designed to help technical communicators understand the importance of seeking an evaluation of the translated work for which they have contracted. The goal is to analyze how translation could be assessed within the paradigm of "pragmatic translation." Assessing translation quality is difficult, and many recognized assessment models exist. These models tend to vary according to the particular criteria being emphasized.

Between September 2004 and January 2006, I conducted a research study in which I analyzed 545 English-to-French translation assignments carried out by university students ranging from the years L2 to M2 (second year of bachelor's degree to final year of master's degree) (Minacori 2010). The analysis had two main objectives: 1) to create an "error tree" useful to teachers (evaluators) and 2) to help students enhance the quality of the translations they produced. The research dealt with fundamental elements to be taken into consideration in translation assessment while it also investigated whether a link could be created between pedagogical and professional translations. Since 2006, I have been testing the assessment model, both with my own students at Université Paris Diderot and with those at other universities (Université de Mulhouse, Université de Besançon, ESIT [Ecole Supérieure d'Interprètes et de Traducteurs, Paris]).

Assessing translation means focusing on errors. What exactly, then, is an "error" in translation? To start, it can be "an act or thought that unintentionally deviates from what is correct, right, or true" (*Concise Oxford* 2005, 459). Martinez Melis and Albir (2001) analyzed different definitions of what constitutes an "error" in the view of different authorities and found the following:

- An error is unintentional, recurring, and systematic according to Spilka, Thomas, and Flamand (1984).
- An error can hamper or deter comprehension or expression as noted by Gile (1992).
- An error will depend on the impact on the final text according to Gouadec (1989). If fatal, the error is deleterious for a translator's reputation.

In my own research on assessment (Minacori 2010), I also investigated several other models relating to errors. This research resulted in the elaboration of an error tree based on five main entries. I have seen the results confirmed in my own practice: These five elements are fundamental for any professional translator. They are as follows:

1. Methodology: Taking into account the mental processes through which translation unfolds. These processes underlie all four of the other elements. Without them, a translator cannot become a professional.
2. Comprehension: The process of understanding a foreign language in any aspect, and the meaning conveyed through that language by an author.
3. Expression: Using one's knowledge of a mother tongue to be able to express the meaning of the author through the process of translation.

4. Research: Required if translators find a gap between the knowledge of the author and their proper knowledge. They will have to conduct research not only on terminology but also on concepts. (For example, for the text on power plants, research could be done on what a steam condenser is.)
5. Revision: The lengthy process through which any text has to go to be published.

These elements constitute a minimum to reach professional quality, but they are time-consuming.

To begin, they represent the reverse of what reading is as a process: Revising goes from concept, via form, to surface elements, while reading involves first decoding surface elements (letters), then forms (words, collocations, phrases) and concepts, and finally meaning. For another, reaching objectivity is difficult when revising one's own translation. This is one of the main reasons why so much time is required for translators to revise a text, yet revision is a compulsory element of any translation. Learning to take the time to see their errors and then revise is one of the main challenges that translation students must meet. Most important for clients to know is that the revision phases remain one of the main challenges that programmers of machine translation software have yet to overcome. Software programs cannot recognize errors that their programmers fail to recognize in their own work.

The error tree (Figure 2.3) is intended to help professional translators assess translations produced translation studies students. For professional translators, the first two columns might appear sufficient when correcting translations. For students, especially those in their first year of study, the third and fourth columns can also provide further explanations to help them better understand their errors. These additional columns make explicit the fundamental outlook underpinning my design of the error tree.

This error tree attempts to clarify for a translator-to-be what constitutes an error, and it can boost revisers' and translation instructors' effectiveness. The tree also reflects an underlying belief that a translation can be judged to be professional if

- a text is clearly understood and correctly expressed,
- research is done into best practices,
- revisions are complete, and
- sound methodologies always underlie the work.

The qualitative and quantitative importance of errors in each category can help revisers perform their jobs and help instructors teach students to perform better.

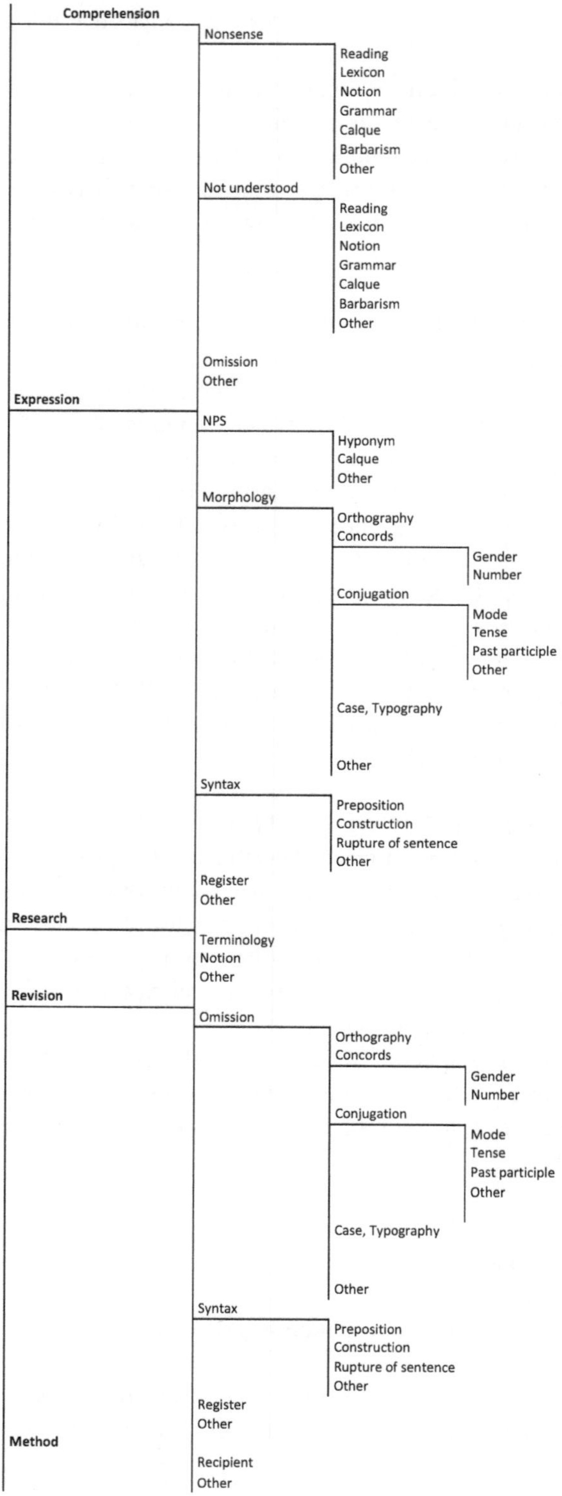

FIGURE 2.3 Error tree stemming from research on 545 translation assessments

Suggested Best Practices for Technical Communicators

This chapter has focused on helping technical communicators overcome commonly held myths and misconceptions about translation so they can work more effectively and in tandem with professional translators to achieve the best possible texts for their clients or recipients reading the texts in translation. These practices are based upon

- inquiries made to organizations in France;
- my research on the technical communication and translation professions, much of it done in conjunction with Trans-Atlantic and Pacific Project collaborations (see www.ndsu.edu/english/trans_atlantic_and_pacific_project/), and in collaboration with University of Limerick and University of Central Florida;
- contacts made with practicing technical communicators who graduated from the University of Paris—Diderot; and
- my role as a course developer in technical communication and in translation studies.

The following list suggests best practices that technical communicators can apply, together with professional translators, to create more feedback and thereby enhance collaboration:

- *Hire a human:* Although those working on artificial intelligence (AI) and applying it to machine translation (MT) are making great strides, they still are stumped when it comes to programming a computer to recognize and then meaningfully render much of what is conveyed in human language. Although many human translators now use MT to speed up their work (for more on this and what is called "post-editing," see Svoboda's chapter, this volume), MT at this point has not advanced far beyond literal translation, with humans still necessary to complete any translation that needs to be reliable (Maylath 2013).
- *Hire a professional:* Untrained bilinguals may claim that they can translate, but you get what you pay for. Would you take a 21st-century computer-equipped car to an untrained backyard mechanic, lacking the car's software codes, for repairs on the car's emissions system? Bilinguals are not automatically translation experts. It takes a long time to become a trained professional translator. For their part, translators turn to trained technical communicators for their expertise in the source text's subject matter and source language's expression.
- *Provide sufficient time:* Translation is a long process; it is never a "press button" procedure. As mentioned, current MT programs, including Google Translate, might be tempting for their speed, as well as their seemingly

low cost. However, they provide users with little more than the gist of a topic, often filled with mistranslations and even gibberish. Again, you get what you pay for. To ensure a highly readable and usable text in translation, human translators need to be involved, and they need adequate time to do a professional job.

- *Exchange information:* Texts by nature include levels of ambiguity and require cultural and experiential inferences. Those levels are heightened when a source text undergoes translation to target texts in target languages. Translators cannot ensure accuracy and precision in meaning without being able to ask questions of the author of the source text in the source language. Communicative isolation permits only guesses—and the guesses will often be faulty. Technical communicators and translators need to be accessible to each other throughout the translation process. This too entails adequate time built into the schedule so that translators can ask their questions and technical communicators can quickly answer them.
- *Learn what's involved:* Already by reading this chapter, you should have an image in mind of the time-consuming labor that professional translators do. However, one chapter can only skim the surface. Reading the other chapters of this book will give you a more complete picture. Even then, though, you will gain a much better view of the challenges presented to translators trying to render your organization's texts into other languages if you talk to the translators about what they are having to cope with.
- *Manage the terminology:* It's incumbent on technical communicators to help the translators organize and harmonize terms, particularly if the translators do not work in-house at your organization. You will be surprised by how many terms familiar to you and your fellow employees raise curious looks from those outside your highly local culture. To facilitate translation, technical communicators and translators need to agree on a list of specialized terms and phrases. Translators can enhance feedback with technical communicators if they point out that a phrase should be modified for the sake of better comprehension, or if they think terms should be harmonized. Translators' revision expertise can be of great help to technical communicators as they create related documentation. In fact, feedback from translators often is what prompts improvements in subsequent editions of the source text, even when those source texts are read by native speakers of the source language.

This listing is not comprehensive, but it represents some of the foundational, and most important, items that technical communicators need to consider when working with translators or creating materials that will be translated.

Conclusions

Translators are experts. Indeed, professional translators have acquired a broad array of competencies involved in the complex processes of the act of translating. They have, for example, multilingual and multicultural knowledge; they are also able to approach a text in its multiple facets. These numerous competencies are the fruits of long training.

It is vital for translators to be included in a continual feedback loop with technical communicators. Doing so is essential to enhance the work of both parties and to create a better working environment for all involved—and to the benefit of the respective employer or client. We have seen that translators call on technical communicators to better understand technical documentation and to use their combined knowledge to focus on the text to be translated. It is thus fundamentally important that technical communicators reciprocate. The ideas presented in this chapter represent a first step in understanding the dynamics essential to fostering such relationships for both the short and the long term.

References

Andréani, Jean-Claude, Françoise Conchon, Jean-Louis Moulin, and Grégoir De Vaissière. 2008. "La Communication de Diversité en Marketing: approche exploratoire." *Management & Avenir* 1, no. 15: 156–173. www.cairn.info/revue-management-et-avenir-2008-1-page-156.htm.

Ballard, Michel. 1999. "Claude-Gaspar Bachet of Mériziac." *Circuit* 65: 20–21.

Ballard, Michel. 2007. *De Cicéron à Benjamin, Traducteurs, Traductions, Réflexions*. Villeneuve d'Ascq, France: Presses Universitaires du Septentrion.

Becker, Rachel. 2014. "Water Repellent Coating Could Make Power Plants Greener." *MIT Technology Review*. November 24, 2014. www.technologyreview.com/news/532756/water-repellent-coating-could-make-power-plants-greener/

Boulanger, Pier Pascale. 2004. "Traduction éthique et société." *TTR: traduction, terminologie, rédaction* 17, no. 2: 57–66.

"Commission statistiques et étude du marché." 2010. *SFT Enquête Tarifs 2009*. Paris: Syndicat national des traducteurs professionnel. www.sft.fr/clients/sft/telechargements/file_front/4c45ab788dee5.pdf

Coirier, Pierre, Daniel Gaonac'h, and Jean-Michel Passerault. 1996. *Psycholinguistique textuelle*. Paris: Armand Colin.

Concise Oxford Dictionary. 2005. Oxford: Clarendon Press.

Durban, Chris. 2011. *Translation: Getting It Right*. Alexandria, VA: American Translators Association.

European Commission. 2015. "Translation: Multilingualism in the European Union." http://ec.europa.eu/dgs/translation/translating/multilingualism/index_en.htm

Gambier, Yves. 2009. "Competences for Professional Translators, Experts in Multilingual and Multimedia Communication." http://ec.europa.eu/dgs/translation/programmes/emt/key_documents/emt_competences_translators_en.pdf

Gile, Daniel. 1992. "Les fautes de traduction, une analyse pédagogique." *Meta* 37, no. 2: 251–262.

Gouadec, Daniel. 1989. Le traducteur, la traduction et l'entreprise. Paris: AFNOR.
Kintsch, Walter. 1974. *The Representation of Meaning in Memory*. New York: Wiley and Sons.
Levý, Jery. 1967. "Translation as a Decision Process." In *To Honor Roman Jakobson: Essays on the Occasion of His Seventieth Birthday* 2, 1171–1182. The Hague and Paris: Mouton.
Martinez Melis, Nicole, and Amparo Hurtado Albir. 2001. "Assessment in Translation Studies: Research Needs." *Meta* 46, no. 2: 272–287.
Maylath, Bruce. 2013. "Current Trends in Translation." *Communication and Language at Work* 1, no. 2: 41–50.
Minacori, Patricia. 2010. "Translation Assessment at University: The Creation of a Computer Interface, Part 1." In *Reconceptualizing LSP: Online Proceedings of the XVII European Language for Specific Purposes Symposium 2009*, edited by Carmen Heine and Jan Engberg. Århus, Denmark. www.asb.dk/fileadmin/www.asb.dk/isek/vibert.pdf
Minacori, Patricia. 2011. "Communicateurs et traducteurs techniques, la logique des échanges." Presentation at the International Conference Echanges, Université Sorbonne Nouvelle, Paris 3, November, 18–19.
Minacori, Patricia, and Lucy Veisblat. 2010. "Technical Translation and Communication: Chicken or Egg?" *Meta* 55, no. 4: 752–768. DOI: 10.7202/045689ar.
PACTE (Procés d'Adquisició de la Competència Traductora i Avaluació). 2003. "Building a Translation Competence Model." In *Triangulating Translation: Perspectives in Process Oriented Research*, edited by Fabio Alves, 43–66. Amsterdam: John Benjamins.
Palumbo, Giuseppe. 2016. "Reviewing Translators' Work." Paper presented at Converging Fields: 1st Conference of the Trans-Atlantic and Pacific Project, Fargo, North Dakota, July 18–20.
Quillard, Geneviève. 1999. "Publicité, traduction et reproduction de la culture." *Babel* 45, no. 1: 39–52. DOI: 10.1075/babel.45.1.04gen
Spilka, Irène, Arlette Thomas, and Jacques Flamand. 1984. *Analyse de traduction: l'universitaire et le praticien*. Ottawa: Editions de l'Université.
Spiro, Rand. 1977. "Remembering Information from Text: The 'State of Schema' Approach." In *Schooling and the Acquisition of Knowledge*, edited by Mark Anderson, Rand Spiro, and William Montague, 137–177. Hillsdale, NJ: Erlbaum.
Van Dijk, Teun, and Walter Kintsch. 1983. *Strategies of Discourse Comprehension*. New York: Academic Press.

AREA 2

WRITING AND CONTENT CREATION

3

CONTROLLED LANGUAGE AND WRITING FOR AN INTERNATIONAL AUDIENCE

Sharon O'Brien

Introduction

Effective technical communication involves having readers in the forefront of one's mind when creating content (Sullivan 1989). This situation is challenging in itself. It becomes more complex when one must consider translation in order to address global audiences. Speaking from a usability perspective, Lentz and Hulst (2000) highlight the challenge:

> Normally in usability studies, one is only dealing with a linguistically homogenous group of native speakers. Usability studies into texts in a multilingual setting, however, require a much more sophisticated approach. In moving toward a more integrated approach, the world of professional communication and document design is becoming increasingly complicated and even more challenging.
>
> *(321)*

The reality is that many individuals will read content *in translation*, either because they have no or limited competencies in the original (i.e., source) language, or because it is their preference to read content in another language. Such situations mean access to information will be mediated by translators. Thus, translators can also be viewed as readers of, and meaning mediators for, the original content.

Translation is a necessity for companies doing business globally, but it represents a significant cost and time investment. To reduce production costs and publication time, natural language processing tools are often used—meaning such tools also have a mediation role in translation. As such, one might expect standard practice to involve collaboration between writing and translation teams. Such collaboration, however, appears to be exceptional, yet many problems encountered during translation could be reduced or eliminated through collaboration and cooperation. This chapter explores the benefits and challenges of writing for translation and, ultimately, for an international audience.

Writing for Translation and Controlled Language

The concept of writing for translation is linked to controlled language (CL), defined by Huijsen (1998) as "an explicitly defined restriction of a natural language that specifies constraints on lexicon, grammar, and style" (2). The objective of a CL is to reduce ambiguity so that language can be more easily "processed" by humans or computers. More specifically, CL aims to improve comprehensibility, readability, intelligibility, and translatability. For example, the CL sentence, "There is no separate import or separate export step as with some other editors," is easier for a computer program to process than the more ambiguous, "There is no separate import or export step as with some other editors," where "separate" is elliptical and understood to modify both of the two nouns that follow it. The CL rule applied here is: "Remove ambiguous scope in coordination." The ambiguous scope of the modifier "separate" has been eliminated by repeating it, arguably giving a computer program more explicit information about the scope of the modifier.

Controlled Language is a term that is not necessarily accepted, or liked, among the technical communication community, primarily due to a dislike that any form of writing would be "controlled."[1] There are several other terms used for such practices, including "standard," "restricted," "simplified" language, and "assisted authoring" or "content optimization." Many terms are vague, and, given that much related earlier research has used "Controlled Language," the contributors to this book use it throughout the volume.

This chapter provides an overview of the topic, along with examples of the historical development of CLs and certain CL rules. The chapter also reviews technologies accompanying the deployment of controlled authoring and the benefits and challenges of CL for technical communication practices. In addition, the chapter provides an overview of writing for machine translation (MT, or automatic translation by computer) as well as future developments in this overall area of translation and technical communication. Through this approach, readers can gain a better understanding of the topic of CL, how it is applicable to MT, and what the future holds for this approach in relation to technical communication.

Overview of the Topic

Huijsen (1998, 2) divides Controlled Languages into two main categories:

- Human-Oriented Controlled Languages (HOCLs)
- Machine-Oriented Controlled Languages (MOCLs)

The objective of a HOCL is to improve comprehension by humans, while that of a MOCL is to improve the processability of language by computer. The difference is described by Huijsen (1998) as follows:

> A general difference is that writing rules for the machine-oriented controlled languages must be precise and computationally tractable—for example, "Do not use sentences of more than 20 words"—while writing rules for human-oriented Controlled Languages may require skills that are presently beyond computers, and they may be somewhat vague—for example, "Make your instructions as specific as possible" and "Present new and complex information slowly."
>
> *(2)*

Essentially, MOCLs seek to craft sentences that make meaning more transparent to computer programs while HOCLs seek to craft sentences that make meaning unambiguous for human processing.

A CL may be destined for *both* human- and machine-processing, in which case the term Dual-Oriented Controlled Language (DOCL) is used. A CL may also be loosely or strictly controlled. In a loosely controlled CL, the rules are akin to general style guidelines, such as the instruction to, "Present new and complex information slowly." A more specific rule (i.e., a stricter control), on the other hand, is exemplified in the following: "Put a time clause at the start of a sentence." CLs are generally composed of rules or guidelines that writers are expected to follow along with a database of "approved terms." This database might also contain alternative terms that individuals should not use.

CLs have been developed and deployed in different domains (e.g., the automotive, aeronautic, heavy machinery, and IT sectors). There is a common general aim across these diverse domains, which is to produce high-quality information that is easily understood by the target audience (i.e., the intended end users) and easily translated. Hidden behind these two general aims is a matrix of more complex considerations that motivate the design and use of CL. Such a matrix would include items and aspects such as

- language competencies of the intended end users (i.e., native versus non-native speakers or readers) and
- whether the content will be translated, by humans or MT, into which languages, and using which kind of MT system.

With regard to the target audience, consideration must be given to whether members of the primary target audience are native speakers of that language. The level of education of the primary target audience and the tasks that they must accomplish with the help of the content must also be taken into account. For example, automotive engineers whose first language is Spanish and who have low levels of competence in English must be able to understand instructions for maintaining and repairing engines if the instructions for doing so are available only in the English language.

Examples of Controlled Languages

A second general goal of CL is that information be *easily translatable*, which prompts questions about underlying objectives. Does "easily translatable" mean by computer (i.e., MT), by humans, or both? What rules should one apply to the source content to ensure translatability? One can understand such factors by examining how CLs are used in different domains.

Caterpillar Fundamental English and Its Off-Shoots

A well-known, early implementation of CL is Caterpillar Fundamental English (CFE), regarded as the predecessor of CLs designed to improve readability and/or translatability. The Caterpillar Corporation introduced CFE in the 1970s, in part due to an increase in the number of products and in the labor costs (Kamprath et al. 1998, 52). Although benefits were realized from its use—for example, a reduction in technical enquiries arising from comprehension problems in the manuals (ibid., 126)—CFE was discontinued after approximately 10 years for technical and cultural reasons. CFE could not adequately describe the increasingly sophisticated products marketed by Caterpillar nor could Caterpillar enforce the basic guidelines. Moreover, due to limited education, service technicians were not learning CFE, and non-native speakers of English cultures wanted documents in their own languages.

Although CFE was discontinued, it was the basis for other CLs including the following:

- E. N. White's ILSAM (International Language for Service and Maintenance) (White 1980)
- Smart's PEP (Plain English Program)
- J. I. Case's Clear and Simple English (CASE)
- Perkin's Approved Clear English (PACE) (Douglas and Hurst 1996, Newton 1992, Huijsen 1998)

Huijsen also credits Smart's PEP with giving birth to other CLs, such as those used by Rockwell International and Hyster (Hyster's Easy Language Program or HELP). ILSAM is reportedly the source for developments such as the AeroSpace and Defence Industries Association of Europe's Simplified English (AECMA SE, now called STE, for Simplified Technical English).

Although Caterpillar abandoned CFE in the early 1980s, developments in writing and publication technology allowed the company to reintroduce CL in the form of Caterpillar Technical English (CTE) in 1991. Development of CTE happened in parallel with development of a CL checker, a tool that automatically checks for compliance with the rule set, and with the rollout of Caterpillar's machine translation system (Kamprath et al. 1998).

AEMCA Simplified English and Its Off-Shoots

AECMA Simplified English provides another example of a successful implementation of controlled language. The safety requirement in the aviation industry provided the impetus for the development and implementation of a CL. Additionally, the need for clarity and accuracy in manuals grew as the number of non-native speaking service technicians increased. AECMA SE was officially renamed "ASD Simplified Technical English™" in January 2005.

AECMA SE was the basis for other CLs in the aviation industry including Boeing Technical English (BTE) (Wojcik and Holmback 1996). Wojcik, Holmback, and Hoard (1998) explain that SE was too restrictive for the technical writers at Boeing and was not well suited to non-procedural sections of maintenance documentation. Hence, Boeing developed its own CL and created a CL checker to support its technical writers.

The automotive sector also uses CL, with BMW and Volvo involved in a CL research and development project called "Multidoc" (Haller 2000). In addition, between 1995 and 2000, General Motors (GM) invested in a CL project called CASL (Controlled Automotive Service Language) (Godden 1998, 2000).

While heavy industries have traditionally been at the forefront of CL research, design, and implementation, application and development has occurred in other sectors like Information Technology (IT). In the 1990s, Sun Microsystems, Nortel, and SAP all invested in CL research and development. Additionally, at the time of this writing, multinational IT companies such as Symantec and Adobe were both using CL to support technical content developers.

Non-English CLs

The impetus to develop CLs has come largely from the English-speaking world, but there have been similar developments in other languages, such as French by Groupe Bull. Moreover, Simplified English served as the basis for the development of Français Rationalisé (FR) developed by "GIFAS" (Groupement des Industries Françaises Aéronautiques et Spatiales) in the 1980s to facilitate the introduction of the concept of SE into the French aerospace industry (Barthe 1998). This group worked backwards to assess which SE rules were applicable to the French language and what new rules were needed. They concluded that it was possible to design one CL to match another, but a "perfect match" was impossible (Barthe 1998, 100). They also concluded that texts written in FR were easier to translate because they were less ambiguous than texts in which FR had not been applied.

There is evidence that the German company Siemens developed a controlled German called Siemens Dokumentationsdeutsch (SDD) (Schachtl 1996).

Other German initiatives include a grammar and style checker for German (Schmidt-Wigger 1998) and a Simplified German initiative (Janssen, Mark, and Dobbert 1996). Efforts by the Swedish company Scania to design and implement a controlled Swedish, which is then translated into several languages, have also been documented (Almqvist and Sågvall Hein 1996, Sågvall Hein 1997, Sågvall Hein, Almqvist, and Starbäck 1997, Sågvall Hein and Almqvist 2000), and Cascales Ruiz and Sutcliffe (2003) mention an initiative aimed at producing CL rules for Spanish. More recently, the sociolinguistics laboratory of Hirosaki University, in Japan, has conducted research on a form of CL known as "Easy Japanese," developed to increase comprehensibility of Japanese by non-native speakers in disaster situations.

Controlled Language Rules

Normally, organizations develop CL rules to meet their specific needs, prioritizing comprehensibility as necessary (as in the case of AECMA Simplified English—see the following paragraphs) or translatability (as with IBM's EEA—see also the following paragraphs). Rule sets are considered to be proprietary, and there is little evidence of openness or sharing.[2] The problems that a CL rule set seeks to solve cannot be terribly dissimilar, and so one might assume that a comparison of CL rules would yield evidence of common rules.

In 2003, I collated and compared eight CL rule sets for English to ascertain what the typical CL rules were and what level of overlap there was across them (O'Brien 2003). The eight CL rule sets included in the analysis were these:

1. AECMA SE (Revision 1, Issue 2)
2. Attempto Controlled English (ACE) (Fuchs and Schwitter 1995, 1996; Schwitter and Fuchs 1996; Fuchs, Schwertel, and Schwitter 1999; Schwertel 2000)
3. Alcatel's COGRAM (Schreurs and Adriaens 1992; Adriaens 1994)
4. IBM's Easy English Analyzer (EEA) (Bernth 1997, 1998a, 1998b, 1999a, 1999b, 2000)
5. GM's Controlled Automotive Service Language (CASL) (Means and Godden 1996, Godden 1998, Godden 2000)
6. Océ's Controlled English (Cremers 2001)
7. Sun Microsystem's Controlled English
8. Avaya's Controlled English

For the purpose of analysis, each rule was classified according to the linguistic feature the rule sought to control leading to the following three categories:

1. Lexical, with a subcategory of lexical-semantic
2. Syntactic
3. Textual, with two sub-categories of Text Structure and Pragmatic

Here we provide an example of each rule category by using the AECMA SE rule set. A lexical rule controls the use of words:

- AECMA SE rule 3.1 is an example of a lexical rule: "Use only those forms of the verb that are listed in the Dictionary."
- SE rule 1.3 is an example of a lexical-semantic rule: "Keep to the approved meaning of the word in the Dictionary. Do not use the word with any other meaning."

A syntactic rule controls the structure of a sentence:

- SE rule 2.1 is an example of a syntactic rule: "Do not make noun clusters of more than three nouns."

A text structure rule specifies how text should be laid out:

- Rule 4.3 is a text structure rule: "Use a tabular layout (vertical layout) for complex texts."
- Rule 6.2 is an example of a pragmatic rule: "Try to vary sentence lengths and constructions to keep the text interesting."[3]

The last rule is not actually advisable when one is writing for translation, because the technologies used to support the translation process are often designed to take advantage of repetitive and consistent structures!

What is noteworthy in Table 3.1 is that some CL rule sets have almost double the number of rules compared to others (e.g., CL4 vs. AECMA SE). This implies that CLs can be more or less controlling. Moreover, it implies that for some CLs the writer would have to obey 60 rules. This explains why technical communicators can harbor a dislike of CL. It also provides motivation for technological support for CLs so that writers do not have to remember so many rules.

Table 3.2 shows the percentage and number (in parentheses) of the types of rules in each CL (i.e., Lexical/Lexical Semantic, Syntactic, and Text Structure/Pragmatic):

TABLE 3.1 Number of rules in each CL

Controlled Language	*Number of Rules*
AECMA SE	60
ACE	36
CL 1	59
CL 2	46
CL 3	35
CL 4	31
CL 5	36
CL 6	38

TABLE 3.2 Number of types of rules in each CL

CL	*Lexical / Lexical Semantic*	*Syntactic*	*Text Structure / Pragmatic*
AECMA	(18) 30%	(12) 20%	(30) 50%
ACE	(11) 30%	(23) 65%	(2) 5%
CL 1	(26) 45%	(24) 40%	(9) 15%
CL 2	(9) 20%	(32) 70%	(5) 10%
CL 3	(15) 45%	(11) 34%	(7) 21%
CL 4	(7) 22%	(13) 42%	(11) 36%
CL 5	(9) 25%	(18) 50%	(9) 25%
CL 6	(17) 45%	(15) 40%	(6) 15%

Syntactic and lexical rules account for the largest proportion of rules overall in the group of CLs analyzed and, with the exception of AECMA SE, text structure and pragmatic rules make up the smallest category for each CL. These figures demonstrate that CLs focus primarily on the control of words and sentence structure and less on specifying how text should be structured. The control of clarity is achieved primarily through lexical and syntactic rules. At the same time, CLs can have different numbers of rule types, suggesting that there is no standard approach to CL implementation.

Only one rule is common to all eight CLs compared here. SE rule 5.1, "Keep procedural sentences as short as possible (20 words maximum)," is echoed in different ways by all CLs where the maximum number of words allowed in a sentence varies from 20, for instructional sentences, to 25, for descriptive ones. Other CLs simply urge the writer not to be too verbose. This implies that the one "rule" upon which all CL designers agree is that sentences should not be "long." This rule is presumably driven by an implicit agreement that longer sentences are more difficult for humans to understand and for machines to process.

Other rules could be said to be common across the CLs. "Common Rules" are defined here as rules that are shared by at least four (i.e., 50%) of the CLs under analysis. The following list details the rules shared by four or more CLs:

- SE rule 1.1, "Use approved words from the Dictionary etc.," is shared by three other CLs.
- SE rule 1.13, "Make your instructions as specific as possible," is shared by three other CLs.
- SE rule 2.1, "Do not make noun clusters of more than three nouns," is shared by five other CLs.
- SE rule 2.3, "When appropriate, use an article (the, a, an) or a demonstrative adjective (this, these) before a noun," is common to six other CLs.

- Six CLs share a rule/recommendation to avoid using gerunds. AECMA SE does not have a specific rule that governs the avoidance of gerunds, but this is covered by rule 3.1, "Only use those verb forms that are listed in the dictionary," and the gerund is not listed in the AECMA SE dictionary.
- SE rule 3.6, "Use the active voice," is shared by six other CLs. The rule takes the form of, "Do not use the passive," (three CLs) to, "Avoid using the passive," in the other three CLs.
- Five CLs share a rule recommending that relative pronouns such as "who," "which," or "that" are not to be omitted. (AECMA SE has no corresponding rule.)

This analysis revealed that only one rule is common to *all* CLs analyzed: to use short sentences. Also, there are only seven rules common to 50% or more of the CLs, and they govern dictionary usage, specificity of information in instructions, noun cluster size, article usage, gerund usage, passive voice, and relative pronoun usage. This analysis suggests that the definition of CLs is largely eclectic and that the linguistic phenomena considered worth controlling vary from organization to organization.

In another analysis, my colleague John Roturier and I compared results for two separate machine translation (MT) experiments for the English-German language pair where CL rules were used (O'Brien and Roturier 2007). Many aspects differed across these two experiments. For example, the MT systems, the source content used, and the evaluation methodologies for the initial experiments all differed. The aim of this analysis was to ascertain whether, despite these differences, certain CL rules presented themselves as having a positive impact on MT output.

A positive impact would present itself in these two studies by reducing the time required in post-MT revision ("post-editing") or in improving the comprehensibility of the MT output, as rated by human evaluators. The conclusions were that some CL rules brought about a positive effect on the MT output across both studies, namely rules governing misspelling, incorrect punctuation, sentence length, and the use of pronouns with ambiguous referents. One can conclude that certain rules seem to contribute to improving human comprehension and lowering post-editing effort. Those rules touch on the basics of good writing: spell correctly, use punctuation correctly, do not write long sentences, and make sure that a pronoun's referent is not ambiguous.

Particularly interesting to notice is that the problems that these CL rules seek to address can be viewed as basic issues in writing: spelling, punctuation, etc. Ensuring that such problems do not exist in the source content can lead to a significant and positive effect on the (machine) translation product. However, fundamental problems, such as typographical errors and poor punctuation, are still frequently reported as being common problems in source texts for translation. Such factors are significant because, for human translation, they place a bigger burden on the translator to disambiguate the source text in order to produce an accurate target text. For machine translation, the processing becomes more complex if language is not well-formed. For example, an MT system will not know how to process an incorrectly spelled word. It will treat it as a word that has not occurred in its lexicon or in its training data. This will most likely result in the transfer of the word as-is to the target sentence and will therefore add to the post-editor's effort.

Controlled Language Technology

CL rule sets vary in terms of the number and type of rules they contain, and there has not been much overlap between rule sets for English. To keep a small set of logical rules in one's head would not be challenging for a technical communicator. Such a small set might include rules governing punctuation, spellings, and the use of the active voice, for example. However, as the rules grow in number and complexity, it is not reasonable to expect writers to remember all rules during the content creation process.

Tools have been designed to assist authors in this task and have names such as "CL Checkers" and "Assisted Authoring Tools." Three current examples are the MAXit Checker, Acrolinx, and HyperSTE. They might come with their own editor or be integrated into a writer's preferred content authoring tool, and they usually come with a set of rules. For example, the MAXit Checker checks for compliance with the ASD-STE100 Simplified Technical English standard. More common, though, is when a tool developer works with a client to create a customized set of rules and integrate the CL checking technology into the client's writing environment.

Once integrated, the tool can be used either interactively or in batch mode by either the writer or a designated editor. In interactive mode, the tool checks for compliance with the rule set as the writer creates content and flags contraventions in real time. Flagging may consist of highlighting the portion of text that contravenes a certain rule, then, when the text is clicked, the rule will pop up indicating what the issue is. The pop-up may also consist of a suggested reformulation of the text, and it may contain an explanation of the issue and some examples of the preferred reformulation. Take, for example, the following sentence: "There is no separate import or export step as with some other editors." One CL checking tool, oriented to machine translation, flags an issue in this sentence as "import or export step" and signals to repeat the modifier in noun phrases that are joined or are linked together. The rule in use here governs "ambiguous scope in coordination." In interactive mode, the sentence would be highlighted and the rule would appear when the writer hovers over the sentence or clicks on the highlighted text.

An alternative to this interactive mode is batch mode, where the writer composes content in the usual manner, and then the draft is run through the CL checker. Text that contravenes rules is highlighted for editing. In such cases, the writer proceeds in a "normal" writing mode, without text highlighting.

When the writer is ready, she can use the CL checker to produce a report of all problematic matters in the text. These matters are loaded into one file, and the writer can then check each one by one. The editing cycle may be performed by the person who created the content or by an editor who is charged with this aspect of the process. It is not uncommon for these tools to also produce a "score," which indicates how compliant a text is with the set of defined rules.

Impact on Professional Communication Practices

For technical communicators, it is easier to write content without adhering to a long list of rules and approved terms. Yet, in contexts where vast volumes of content are created to describe complex systems and multiple communicators contribute to that content, nobody can deny the benefits of using a set of guidelines. The guidelines ensure that the content has a consistent look and feel and that it is comprehensible and usable by the intended readers. Moreover, in certain contexts, comprehensibility is essential. General style guides contribute significantly in meeting such technical communication requirements, though their generality means that they sometimes fall short of being adequate, in which case more precise rules become necessary. Further, where content is translated into many target languages, having a consistent and unambiguous starting point for translation is undeniably beneficial.

Benefits of CL

The benefits of CL fall into two categories:

1. Matters centered around the original text
2. Matters centered around the target, or translated, text in a context where that text is translated

The key words associated with the first category include readability and comprehensibility, while the term translatability is associated with the second category.

Readability

Means and Godden (1996) define "readability" as "a factor of the complexity of sentence structures, the amount of ambiguity, and the use of standardized vocabulary" (107). If a document has a high readability factor, it likely has sentences with simple linguistic structures open to only one interpretation. How to best measure readability has been the focus of many studies and is still a contested topic. As Atwater (1998) notes, a number of automatic readability measures exist, such as the Flesch-Kincaid, Coleman-Liau, and Bormuth grades. These readability indices count the number of syllables, words, or characters per sentence and the length of the sentence. Nonetheless, it is a topic that prevails in discussions of content creation and translation. It is somewhat taken for granted that, by reducing the complexity and ambiguity in content through the application of CL rules, we ultimately increase readability and comprehensibility. However, readability researchers have demonstrated that readability is a property of the reader, not just of the text, and that it is dependent on level of education, of expertise, and on reader type (O'Brien 2010).

Although there are questions about the use of readability measures, it is still interesting to see how revising a text impacts these scores. As mentioned, two standard readability measures are the Flesch-Kincaid Reading Ease and the Flesch-Kincaid Grade Level scores. Using measures involving the total number of words and the total number of syllables, the first gives a score of between 0 and 100. The higher the score, the more readable the sentence or text. The second score (Flesch-Kincaid Grade Level) gives an indication of the grade level in the American education system that a reader must have reached in order to understand the text. For this measure, a score over 22 indicates that the reader should be at graduate level. Tables 3.3 and 3.4 show these two scores for a sentence and then how the scores change when the sentence is rewritten in shorter chunks:

TABLE 3.3 Readability scores before revision

Sentence	*Flesch-Kincaid Reading Ease Score*	*Flesch-Kincaid Grade Level*
This causes bullets to appear in unordered lists, numbers and letters to appear in ordered lists, and other formatting items that the Xyvsion, XHTML, or HTML formatters introduce.	41.6	15.2

TABLE 3.4 Readability scores after revision

Sentence	*Flesch-Kincaid Reading Ease Score*	*Flesch-Kincaid Grade Level*
This causes bullets to appear in unordered lists.	71.8	5.2
It causes numbers and letters to appear in ordered lists.	78.2	4.8
Other formatting items that the Xyvsion, XHTML, or HTML formatters introduce also appear in lists.	50.6	9.9

The examples in Tables 3.3 and 3.4 clearly show that the readability score can be improved for one sentence by making it shorter.

The benefit of a high readability factor is increased comprehension—of utmost importance when the document's aim is to instruct the user on how to perform tasks on technical equipment (e.g., ensure equipment functions properly). Ease of comprehension is especially important when the user is a non-native speaker of the source language.

Translatability

"Translatability" means how easy it is to translate content. As with "readability," the concept is more complex and linked to the competencies of the person doing the translation (for human translation) or the sophistication and suitability of the engine (with MT). Means and Godden (1996) link translatability to the readability of content by stating "translatability is a factor of the readability level, and the number of words" (107). Thus, the more translatability indicators that we can remove, the easier the text should be to translate.

If content is more translatable with the application of CL rules, we can assume that this would also lead to increased productivity in translation, since translators would have to spend less time disambiguating problematic constructions. In an era when it is expected that translated content is published simultaneously with the original content, translation productivity is of key importance. If we accept that there is a link between readability of a source text and translatability, this is an advantage of CL. However, this link is contested, and recent developments in MT technology have further complicated the equation.

Quality

Using CL does not automatically guarantee a high level of quality in content, but it is an expected benefit. Of course, the concept of "quality" is particularly difficult to quantify, and organizations have different criteria for measuring quality. For example, in the aeronautic industry, documentation quality could be measured by the number of clarifications required by a service technician before a maintenance task can be correctly executed. In the automotive sector, it might be measured according to the number and type of errors made according to the Society of Automotive Engineer (SAE)'s translation quality standard—the J2450 (Woyde 2001).

By comparison, the quality of a software manual could be measured according to the ease with which required information can be found and followed (Byrne 2004). Generally, CLs are tailored towards the specific goals of the user. In the case of GM, the aims were to increase the quality of source documentation, leading to better service and, ultimately, lower costs, but there was the added benefit of complying with language laws in certain regions:

> Pertaining specifically to translation of service manuals, the following additional benefits are expected:
>
> - reduced lead time in producing translations;
> - compliance with language laws in Quebec and other countries (*sic*).
>
> *(Means and Godden 1996, 109)*

Whether an organization is trading in the heavy machinery or the IT sector, the challenges of commercial survival are similar: meet your customers' demands for new and better functionality, or they will purchase elsewhere. This concept equates to a requirement for new or updated models or versions on a frequent basis. When a new feature is introduced, it must be documented, and the information must then be translated. However, this information is not distributed in isolation but is instead added to existing information and to entire manuals, usually through content management systems. Therefore, the retrievability of previously published information becomes important.

When a technical communicator updates content, there may be a temptation to improve or edit previously written phrases, sentences, or paragraphs. Obviously, if an update is necessary due to a change or addition of a feature, this makes perfect sense. Technical communicators working in monolingual environments, however, are frequently unaware of the cost of minor, nonessential, or preferential improvements. For example, if a writer decides to change how information has been expressed from one version of a document to the other, this change will have a knock-on effect on the cost of translation, as explained next.

Typically, large-volume translation scenarios make use of translation memories (TM), where every segment previously translated is stored in the TM (a database) along with the translated segment previously produced. Sentences that were previously translated are leveraged from the TM, and translators are paid according to how similar the new sentence is to the one that exists in the TM. Therefore, every punctuation mark or word changed by the professional communicator results in increased translation costs. Further, the more target languages into which the information is due to be published, the higher the costs.

Using CL to control consistency and style can have a positive effect on translation costs. The combination of content management systems with TM, CL and MT presents a powerful solution for the retrievability and recycling of information and is a solution that has been implemented by some companies since the 1990s. By being aware of how CL can positively impact on translation, professional communicators can save undue costs for their employers or clients.

The technology that has been developed to support the process of writing for translation has become essential for using CL. Without this technology it would be very difficult for writers to apply a long list of rules to the content that they have created. Moreover, the proper use of terminology databases in conjunction with CL technology ensures that terminology is consistent. In addition, the tools' automatic checking and, in some cases, functionality of automatic reformulation make compliance easier. The technology also introduces considerable challenges, which I discuss in the following section, along with the general challenges that come with CLs.

Challenges of CL

The introduction of a tightly controlled writing and translation environment presents both drawbacks and benefits. The main challenges are explored in this section.

Cost

Although I mentioned that a reduction in costs, especially those associated with translation, is a direct benefit of CL, it is also one of the challenges, as the implementation of a CL is not cost-free. Implementation might involve definition and creation of a CL. It might also involve development or implementation and customization of the associated technology, training, and system maintenance. Understandably, any organization considering the introduction of CL has to be convinced of the return on investment (ROI) before embarking on such a time- and resource-demanding project, and proving ROI is the challenge faced by CL technology solution providers. Some of the benefits of high-quality translation include

- fewer accidents;
- reduced requests for technical support;
- increased readability, clarity, or accuracy leading to increased customer satisfaction; and
- increased translatability, productivity, and quality in translation.

For many organizations, failure to enunciate these benefits has caused them to abandon their CL initiatives.

Learning Curve

The personnel most likely affected by CL are technical communicators, who may be experts in specific fields (e.g., mechanical engineering) as well as expert communicators, but they do not necessarily have the linguistic training that is sometimes required for the implementation of specific CL rules. Consider the following sentences: "Expand post-nominal modifiers into full relative clauses," or, "Definite noun phrases used anaphorically always refer to the most recent accessible noun phrase that is suitable, i.e., that has the same noun and at least the same adjectives, possessive noun, *of*-prepositional phrase, and apposition as the referring definite noun phrase." As these examples reveal, writers might have to negotiate a learning curve when CL is introduced.

Resistance

Introducing CL can be viewed as a threat to the very people who are expected to embrace it. Technical communicators may view CL as a threat because they are being asked to unlearn their writing strategies and to relearn new ways of writing. Writing in a CL is difficult because demands of conformity and consistency are put on the writer. This can be interpreted as a clasp on the creativity of the technical communicator. Also, new technology, in the form of a CL checker, must be learned and applied. Nevertheless, resistance can be overcome. Reuther (1998), reporting on the degree of skepticism and the fear of restriction displayed by authors who were being asked to use new CL technology, explains how this fear was overcome:

> . . . the technical authors . . . became aware that the tools they are intended to use will not dictate to them in an inflexible and arbitrary way a certain style of writing, but that the tools support them by merely giving indications where possibly a correction or reformulation might contribute to more clarity and readability in their documents.
>
> *(180)*

Illustrating how CL and related technologies helps vs. hinders writers can prompt acceptance of new ways of working. It will, however, take time to become part of the normal process.

Another way of increasing acceptance is by involving technical communicators in the CL initiative from the beginning and by accommodating their existing work practices, where possible. Interestingly, GM seemed to learn this over a period of time when introducing CASL into its organization. In 1996, GM's intention was that CASL "...will impose new restrictions on the work of authors and translators, and will drive some changes in their work patterns" (Means and Godden 1996, 110). However, by 2000, the company's approach had changed significantly: "We realized that it was essential for success to minimize the impact of CASL technology on the authoring community" (Godden 2000, 15).

Time

Introducing CL into the content production cycle means lengthening the cycle and making it more complex by adding the extra steps of CL checking and revision. The advantage of using a dedicated editor is that the impact on the author's writing processes is reduced. Authors simply have to implement the CL editor's corrections. The disadvantage of this approach is that the authors have less incentive to learn the CL rules and are more removed from the technology. Also, having numerous authors and only one or two CL editors means that a human bottleneck can exist in the content production cycle.

Process

Introducing new technology into preexisting practices is never without struggles. With CL technology, it is not just a question of using new technology, but also of changing how content is created. In interactive mode, notifications will pop up indicating that text that the writer has just created does not comply with one or more of the CL rules. This can interrupt the flow of thought necessary for creating the content and therefore could be irritating for the writer.

As with all natural language-processing technology, precision and recall are challenging: Not all "flags" might be accurate and not all "errors" might be detected. The remaining challenge that is worth mentioning here is that CL checkers force writers to take responsibility not just for the comprehensibility of their own content but also for the translatability of it. This responsibility is something that they perhaps did not have to be overly concerned about prior to the introduction of CL.

Writing for Machine Translatability

An important connection exists between CL and machine translatability. Given the assumption that CL improves translatability and specifically machine translatability, it is often introduced at the same time as MT.

Research conducted in the past indicates that the quality of MT output can be significantly improved when input is controlled and when CL rules are first applied to the source texts being translated (Adriaens et al. 1996, Mitamura et al. 1998, Adriaens et al. 2000, EAMT and CLAW 2003). This research was conducted when the dominant approach to MT was "rules-based," i.e., the MT engine was developed by encoding specific linguistic rules for source languages, target languages, and how particular linguistic constructions should be transferred between specific source and target languages. Such MT engines, commonly referred to as RBMT (rules-based machine translation), relied heavily on dictionaries of both the source languages and target languages. These dictionaries would be compiled and maintained by linguists.

Previous research revealed source text written according to specific CL rules could improve the quality of MT output from an RBMT system. This stands to reason: If the source text is tailored to suit how the MT system processes the source language, a better-quality translation should emerge. I took this research one step further (O'Brien 2006) by investigating what impact specific CL rules would have on the task of post-editing (i.e., the editing of MT output to render it into a form that was acceptable for publication). My experiment involved measuring the post-editing effort for professional German language translators, where I correlated effort with specific English linguistic features to see which of those features most negatively affected productivity. Features such as the use of the *-ing* form (e.g., "Ending tags point the arrow to the left."), ungrammatical constructs (e.g., "The editor work direct with XML files"), the use of "(s)" to mark an alternative plural form (e.g., "From the Windows Explorer, double-click on a document(s) or entity"), non-finite verbs (e.g., "Figure 12 shows the default display with the tags shown"), long noun phrases (e.g., "Completeness Menu Item"), and short segments (four words or fewer, such as "Select C drive") were found to have the highest negative impact on post-editing productivity. CL rules that control these features in an English source text could therefore have a positive impact on MT output.

Since these early studies, a significant development has taken place in MT research and development, namely a move away from the RBMT-type system towards data-driven, statistical systems known as statistical machine translation, or SMT (Koehn 2010). Put simply, an SMT system makes use of corpora to build an MT system that calculates the probabilities of words and phrases being translated in specific ways in the target language. This approach differs greatly from the rules-based approach. Hence, a new question has emerged asking about the appropriateness of controlled language rules for SMT systems.

What is important in the probabilistic, statistical approach is the similarity of the content for translation to the content that was used to build the system. For example, if the content used to build (or "train") a system was in the form of technical instructions, and if the content submitted to the system for translation is web marketing content, then it is probable that the differences between the two content types would result in a poorer performance by the MT system. Similarly, if the content used to train the system has not been authored using CL rules, but the content submitted for translation has been authored using CL rules, then there may be an impact on the quality of the MT output.

Doherty (2012) addressed this issue by using CL-authored and non-CL-authored content and then measuring the subsequent readability and comprehensibility of the MT output. He found higher levels of readability and comprehensibility for the MT output for the content that had been subjected to CL rules, even though the MT system had not been built using such content. This finding suggests that the use of CL can have a positive impact on SMT-generated output, even if the SMT system has not been built using such content.

However, this intriguing discovery needs considerably more research in the future. We can at least tentatively conclude that using some form of writing for machine translation rules can contribute positively to the quality of MT output and to post-editing productivity.

Projections for the Future

Changes in existing processes are not easily accepted in organizations, and the introduction of CL is no exception. What is of interest here are the parallels between the professional communication community and the professional translation community. Over the past decades, roles in both have become increasingly technologized. In the 1990s, translators had to navigate the game-changing introduction of TM technology, which radically altered work practices, pricing models, and expectations around productivity. Today, with exceptional advances in MT technology, at least part of the professional translation community is undergoing another radical change with a move to the revising and editing of MT output. This development puts new demands on the pricing models and on translator productivity. Translators are intelligent and creative professionals, and some resist new technologies and radical changes—sometimes for good reasons, sometimes because humans just do not always embrace change readily. Despite these parallels between the two communities and the benefits of linking the content production cycle with the content translation cycle, few organizations have managed to successfully implement a joined-up solution.

A new challenge has emerged in technical communication and translation settings: user-generated content (UGC). Users are now generating vast quantities of content in myriad domains, including professional communication settings. It is not only charitable or socially advantageous projects that are benefitting from this movement. Even IT multinationals are taking advantage of the valuable expertise of their users and of the content they generate online. This user-generated content is created in multiple languages, but there is often a dominant language, such as English or French, in which most content is generated. The problem arises in how to translate this knowledge into other languages. Frequently the solution is MT, because companies do not have a budget for such large-scale human translation endeavors.

The use of MT for UGC introduces a complication, however, as UGC is sometimes not of the same high quality as professionally written content. Punctuation is often missing or misapplied, spellings can be incorrect, typographical errors and abbreviations occur frequently. MT systems do not deal well with such ill-formed content; thus, there is an even greater need for writing for translation solutions.

The most desirable solution would be for technology, in the form of (semi-) automatic application of correction and CL rules, to come to the rescue. For example, Gerlach et al. (2013) investigated the effects on SMT output of pre-editing UGC, written in French and machine-translated into English. The UGC presented an additional layer of complexity, as the source text content contained many contraventions of grammar rules (e.g., typographical errors, lack of punctuation, infringement of basic grammar rules). They found, by pre-editing the content with CL rules, that the post-editing time was reduced for many of their sample sentences and there was no loss of quality linked with this time gain. (Semi-) automatic pre-editing of UGC is certainly an area where CL technology might be deftly applied in the future.

Conclusion

This chapter has provided an overview of initiatives dedicated to creating content comprehensible to non-native speakers of the authoring language. It has also discussed the creation of content that is translatable by human translators using translation memory tools or by post-editing machine translated output. Research suggests that writing with translation in mind can have benefits in terms of translation quality and productivity in the translation cycle. At the same time, the introduction of such initiatives is not without its challenges, not least of which is convincing source-content authors that their audience extends beyond just those who will read their source content but also to translators—and eventually, through their work, to an international readership.

Notes

1 Evidence for this comes mainly from industry commentaries on CL, e.g., two blog posts from large language service providers in 2010 (Lionbridge: http://blog.lionbridge.com/translation/2010/06/07/controlled-language-benefits-document-translation/) and 2013 (Moravia: http://info.moravia.com/blog/bid/313981/Watch-Your-Language-3-Reasons-Writers-Dislike-Controlled-Language).

2 Between 2001 and 2006, I was involved in CL research in which companies agreed to share CL rules for research purposes. However, several companies insisted on a nondisclosure agreement. While STE, for example, is available to those who may wish to use it, one had to pay for access, so there is still little evidence of sharing or open-sourcing of CL rule sets at the time of this writing.

3 The classification of rules is by no means unproblematic, but a full discussion is beyond our scope here. For a more detailed discussion of this problem, see O'Brien 2003.

References

Adriaens, Geert. 1994. "Simplified English Grammar and Style Correction in an MT Framework: The LRE SECC Project." Paper presented at the Translating and the Computer 16 Conference, London, November 10–11.

Adriaens, Geert, Jeffrey Allen, Arendse Bernth, Kurt Godden, Teruko Mitamura, Eric Nyberg, Rick Wojcik, and Remi Zajac, eds. 2000. *Proceedings of the 3rd International Workshop on Controlled Language Applications*. Seattle, WA.

Adriaens, Geert, Roger Havenith, Rick Wojcik, and Bruno Tersago, eds. 1996. *Proceedings of the 1st International Workshop on Controlled Language Applications*. Leuven, Belgium: Centre for Computational Linguistics.

Almqvist, Ingrid, and Anna Sågvall Hein. 1996. "Defining Scania Swedish: A Controlled Language for Truck Maintenance." Paper presented at the 1st International Workshop on Controlled Language Applications, Centre for Computational Linguistics, Leuven, Belgium, March 26–27.

Atwater, Kathleen. 1998. *Nortel Standard English as a Quality and Reliability Tool: Distributed Report*. Ottawa, ON, Canada: Public Carrier Networks Information Development, Nortel.

Barthe, Kathy. 1998. "GIFAS Rationalised French: Designing One Controlled Language to Match Another." Paper presented at the 2nd International Workshop on Controlled Language Applications, Language Technologies Institute, Carnegie Mellon University, Pittsburgh, PA, May 21–22.

Bernth, Arendse. 1997. "EasyEnglish: A Tool for Improving Document Quality." Paper presented at the 5th Conference on Applied Natural Language Processing, Washington, DC, March 31–April 3.

———. 1998a. "EasyEnglish: Addressing Structural Ambiguity." Paper presented at the 3rd Conference of the Association of Machine Translation in the Americas, Langhorne, PA, October 28–31.

———. 1998b. "EasyEnglish: Preprocessing for MT." Paper presented at the 2nd International Workshop on Controlled Language Applications, Language Technologies Institute, Carnegie Mellon University, Pittsburgh, PA, May 21–22.

———. 1999a. "Controlling Input and Output of MT for Greater User Acceptance." Paper presented at the Translating and the Computer Conference 21, London, November 10–11.

———. 1999b. "EasyEnglish: A Confidence Index for MT." Paper presented at the 8th International Conference on Theoretical and Methodological Issues in Machine Translation, Chester College, England, August 23–25.

———. 2000. "EasyEnglish: Grammar Checking for Non-Native Speakers." Paper presented at the 3rd International Workshop on Controlled Language Applications, Seattle, WA, April 29–30.

Byrne, Jody. 2004. "Textual Cognetics and the Role of Iconic Linkage in Software User Guides." Ph.D. diss., Dublin City University.

Cascales Ruiz, Remedios, and Richard Sutcliffe. 2003. "A Specification and Validating Parser for Simplified Technical Spanish." Paper presented at the Joint Conference Combining the 8th International Workshop of the European Association for Machine Translation and the 4th International Workshop on Controlled Language Applications, Dublin City University, Ireland, May 15–17.

Cremers, Lou. 2001. "Towards an Automated Translation Workflow at Océ Technologies." *International Journal for Language and Documentation* 9: 24–25.

Doherty, Stephen. 2012. "Investigating the Effects of Controlled Language on the Reading and Comprehension of Machine Translated Texts: A Mixed-Methods Approach." Ph.D. diss., Dublin City University.

Douglas, Shona, and Matthew Hurst. 1996. "Controlled Language Support for Perkins Approved Clear English (PACE)." Paper presented at the 1st International Workshop on Controlled Language Applications, Centre for Computational Linguistics, Leuven, Belgium, March 26–27.

EAMT (European Association for Machine Translation) and CLAW (Controlled Language Applications Workshop). 2003. *Controlled Language Translation: Proceedings of the Joint Conference Combining the 8th International Workshop of the European Association for Machine Translation and the 4th International Workshop on Controlled Language Applications*. www.mt-archive.info/00/CLT-2003-TOC.htm

Fuchs, Norbert, and Rolf Schwitter. 1995. "Attempto Controlled Natural Language for Requirements Specifications." Paper presented at the 7th IIPS 95 Workshop on Logic Programming Environments, Portland, OR, January 13–16.

Fuchs, Norbert, and Rolf Schwitter. 1996. "Attempto Controlled English (ACE)." Paper presented at the 1st International Workshop on Controlled Language Applications, Centre for Computational Linguistics, Leuven, Belgium, March 26–27.

Fuchs, Norbert, Uta Schwertel, and Rolf Schwitter. 1999. *Attempto Controlled English Language Manual Version 3.0*. Geneva, Switzerland: Institut für Informatik der Universität Zürich.

Gerlach, Johanna, Victoria Porro, Pierrette Bouillon, and Sabine Lehmann. 2013. "Combining Pre-Editing and Post-Editing to Improve SMT of User-Generated Output." Paper presented at the 2nd Workshop on Post-Editing Technologies and Practice, 14th Machine Translation Summit, Nice, France, September 2–6.

Godden, Kurt. 1998. "Controlling the Business Environment for Controlled Language." Paper presented at the 2nd International Workshop on Controlled Language Applications, Language Technologies Institute, Carnegie Mellon University, Pittsburgh, PA, May 21–22.

———. 2000. "The Evolution of CASL Controlled Authoring at General Motors." Paper presented at the 3rd International Workshop on Controlled Language Applications, Seattle, WA, April 29–30.

Haller, Johann. 2000. "MULTIDOC: Authoring Aids for Multilingual Technical Documentation." Paper presented as the 1st Congress of Specialized Translation, Barcelona, March 2–4.

Huijsen, Willem-Olaf. 1998. "Controlled Language: An Introduction." Paper presented at the 2nd International Workshop on Controlled Language Applications, Language Technologies Institute, Carnegie Mellon University, Pittsburgh, PA, May 21–22.

Janssen, Gerd, Gerhard Mark, and Bernd Dobbert. 1996. "Simplified German: A Practical Approach to Documentation and Translation." Paper presented at the 1st International Workshop on Controlled Language Applications, Centre for Computational Linguistics, Leuven, Belgium, March 26–27.

Kamprath, Christine, Eric Adolphson, Teruko Mitamura, and Eric Nyberg. 1998. "Controlled Language for Multilingual Document Production: Experience with Caterpillar Technical English." Paper presented at the 2nd International Workshop on Controlled Language Applications, Language Technologies Institute, Carnegie Mellon University, Pittsburgh, May 21–22.

Koehn, Philipp. 2010. *Statistical Machine Translation*. Cambridge: Cambridge University Press.

Lentz, Leo, and Jacqueline Hulst. 2000. "Babel in Document Design: The Evaluation of Multilingual Texts." *IEEE Transactions on Professional Communication* 43, no. 3: 313–322.

Mitamura, Teruko, Eric Nyberg, Geert Adriaens, Linda Schmandt, Rick Wojcik, and Remi Zajac, eds. 1998. *Proceedings of the 2nd International Workshop on Controlled Language Applications*. Pittsburgh, PA: Language Technologies Institute, Carnegie Mellon University.

Means, Linda, and Kurt Godden. 1996. "The Controlled Automotive Service Language (CASL) Project." Paper presented at the 1st International Workshop on Controlled Language Applications, Centre for Computational Linguistics, Leuven, Belgium, March 26–27.

Newton, John. 1992. "The Perkins Experience." In *Computers and Translation: A Practical Appraisal*, edited by John Newton, 46–57. London and New York: Routledge.

O'Brien, Sharon. 2003. "Controlling Controlled English: An Analysis of Several Controlled Language Rule Sets." Paper presented at the Joint Conference combining the 8th International Workshop of the European Association for Machine Translation and the 4th International Workshop on Controlled Language Applications, Dublin City University, Ireland, May 15–17.

———. 2006. "Machine-Translatability and Post-Editing Effort: An Empirical Study Using Translog and Choice Network Analysis." Ph.D. diss, Dublin City University.

———. 2010. "Controlled Language and Readability." In *Translation and Cognition.* American Translators Association Scholarly Monograph Series, edited by Gregory M. Shreve and Erik Angelone, 143–165.. Amsterdam/Philadelphia: John Benjamins.

O'Brien, Sharon, and Johann Roturier. 2007. "How Portable Are Controlled Language Rules? A Comparison of Two Empirical MT Studies." Paper presented at the 11th Machine Translation Summit, Copenhagen Business School, Denmark, September 10–14.

Reuther, Ursula. 1998. "Controlling Language in an Industrial Application." Paper presented at the 2nd International Workshop on Controlled Language Applications Workshop, Language Technologies Institute, Carnegie Mellon University, Pittsburgh, PA, May 21–22.

Sågvall Hein, Anna. 1997. "Language Control and Machine Translation." Paper presented at the 7th International Conference on Theoretical and Methodological Issues in MT, Santa Fe, NM, July 23–25.

Sågvall Hein, Anna, and Ingrid Almqvist. 2000. "A Language Checker of Controlled Language and Its Integration into a Documentation and Translation Workflow." Paper presented at the Translating and the Computer 22 Conference, London, November 16–17.

Sågvall Hein, Anna, Ingrid Almqvist, and Per Starbäck. 1997. "ScaniaSwedish: A Basis for Multilingual Machine Translation." Paper presented at the Translating and the Computer 19 Conference, London, November 13–14.

Schachtl, Stefanie. 1996. "Requirements for Controlled German in Industrial Applications." Paper presented at the 1st International Workshop on Controlled Language Applications, Centre for Computational Linguistics, Leuven, Belgium, March 26–27.

Schmidt-Wigger, Antje. 1998. "Grammar and Style Checking for German." Paper presented at the 2nd International Workshop on Controlled Language Applications, Language Technologies Institute, Carnegie Mellon University, Pittsburgh, PA, May 21–22.

Schreurs, Dirk, and Geert Adriaens. 1992. "From COGRAM to ALCOGRAM: Toward a Controlled English Grammar Checker." Paper presented at the 14th International Conference on Computational Linguistics, Nantes, France, August 23–28.

Schwertel, Uta. 2000. Controlling Plural Ambiguities in Attempto Controlled English (ACE)." Paper presented at the 3rd International Workshop on Controlled Language Applications, Seattle, WA, April 29–30.

Schwitter, Rolf, and Norbert Fuchs. 1996. "Attempto: From Specifications in Controlled Natural Language Towards Executable Specifications." Paper presented at the EMISA Workshop Natürlicher Entwurf von Informationssystemen, Akademie Tutzing, Germany, May 28–30.

Sullivan, Patricia. 1989. "Beyond a Narrow Conception of Usability Testing." *IEEE Transactions on Professional Communication* 32, no. 4: 256–264.

White, E. N. 1980. "Using Controlled Languages for Effective Communication." Paper presented at the 27th International Technical Communication Conference, Society for Technical Communication, Washington, DC, May 14–17.

Wojcik, Rick, and Heather Holmback. 1996. "Getting a Controlled Language off the Ground at Boeing." Paper presented at the 1st International Workshop on Controlled Language Applications, Centre for Computational Linguistics, Leuven, Belgium, March 26–27.

Wojcik, Rick, Heather Holmback, and James Hoard. 1998. "Boeing Technical English: An Extension of AECMA SE beyond the Aircraft Maintenance Domain." Paper presented at the 2nd International Workshop on Controlled Language Applications, Language Technologies Institute, Carnegie Mellon University, Pittsburgh, PA, May 21–22.

Woyde, Rick. 2001. "Introduction to the SAE J2450 Translation Quality Metric." *Language International*, April, 37–39.

4

QUALITY IN TRANSLATION THROUGH "CONTROLLED" WRITING, EDITING, AND/OR REVISING

Maria Teresa Musacchio

Introduction

Historically, translation was usually done in isolation at home or in small companies with translators occasionally visiting clients and libraries to learn more about the tasks and topics at hand. Computing—and especially the internet—now allows translators to become part of virtual teams. They are more integrated into the workflow of document production and can more easily liaise with authors, clients, and fellow translators for discussion and advice.

As a consequence, translating is now a (tele)collaboration activity within international communication. This situation means that writing, editing, and revising for translation have changed. As Scarpa and Minacori point out elsewhere in this volume, writers now need to account for special requirements of documentation intended for an international readership and hence adapt their texts accordingly. Translators, in turn, have become acutely aware of their role as mediators in international communication, both as writers and as revisers/editors of documents to meet criteria for quality and comply with standards for quality (Musacchio 2006). Thus, writing, with attendant editing and revising, is central both to the production of the original text and to its translation: It therefore needs to be "controlled" (i.e., subject to "an explicitly defined restriction of a natural language that specifies constraints on lexicon, grammar, and style" [Huijsen, 1998, 2; see also O'Brien, this volume]). This process produces a "style" that takes account of factors that improve or hinder localization and internationalization of documents in a globalized world (see O'Brien, this volume).

This chapter provides an overview of such ideas for readers new to translation, localization, and internationalization. It introduces strategies for writing, revising, and/or editing to produce high quality translations for an increasingly globalized world. It explores how to write, revise, and edit while focusing on professional translators' needs, and it does so in a professional context requiring increasingly rapid translations and examines how to maintain translation quality within such contexts.

Overview

In the 1950s, there was a significant increase in translation as quicker communication and transport systems made international contacts and global trade more frequent. The translation of scientific and technical documentation increased exponentially, and translators no longer considered themselves craftspeople but as professionals who made a living translating (for the details of how this came to be, see Minacori's chapter in this volume). Several international organizations and large companies started recruiting staff to create in-house translation services, and language professionals established agencies to provide services—translating, interpreting, and language training—in several languages. Development in the language industry was largely driven by the idea that computers would soon handle human languages automatically, and fully automated machine translation would soon be a reality. But natural languages—languages that have developed as the native speech of people—proved more difficult to tackle than expected. Thwarted ambitions made for more realistic objectives as computer-assisted translation (CAT) tools developed. Word processing software and desktop publishing became widespread. In time, software was produced to help translators manage terminology and create databases of their translations. Translators could now retrieve, reuse, and adjust previous work to suit the requirements of new translations and new clients. These tools, often called translation memories (TMs), are now essential skills for any professional translator (Sager 1994).

A second change occurred in the late 1980s, when software firms started demanding more and faster translations of their products. Documentation had to be "localized" to fit the target culture. In response, language service providers (LSPs) developed and expanded, all the while looking at other sectors to identify practices that would make translation flows and management smoother. As a consequence, manufacturing and managerial practices were introduced to ensure efficiency.[1]

A new outlook on the industry and new, more powerful technological tools have had an effect on writing in and for translating, substantially altering views on translation quality. Technical communicators need to draft documents—the source texts for translations—with the idea they will likely be translated. According to the European Union's 2006 machinery directive, for example, "All machinery must be accompanied by instructions in the official Community language or languages of the Member State in which it is placed on the market and/or put into service" (European Parliament and Council of the European Union 2006). As a consequence,

- instructions accompanying the machinery must be "original instructions" or a "translation of the original instructions," in which case the translation must be accompanied by the original instructions;
- translators need to keep pace with developments in technology and devise strategies that allow them to meet increasing demand for faster and faster output while ensuring high quality translations; quality is a concept enshrined in industry standards which provide guidelines to streamline workflows and ensure that good products are turned out (Bass 2013); and
- quality implies that documents have to speak the language of the target audience and should be similar to texts written within the target language community and in that special domain of the discipline (Wright 1993, 70).

This chapter examines these factors by focusing on the following topics: controlled writing, editing, and/or revising to ensure high quality translations.

"Controlled" Writing

Writing for translation is a form of controlled writing. Generally speaking,

> A **controlled language** is a subset of a natural language which is created using various rules to restrict the grammar and vocabulary which can be used in a text. The purpose of a CL is to reduce ambiguity and complexity in texts and it does this by limiting the ways in which information can be expressed.
>
> *(Byrne 2012, 142)*

"Controlled" here means documents are drafted by taking account of requirements emerging as a result of

- internationalization or creating products in such a way that they can be quickly and easily adapted to different local conditions and
- glocalization or designing materials that can be distributed globally but also work in different local contexts.

Consequently, if writers and translators are constantly in touch and discuss matters with translators during document preparation and coauthoring of texts, the final quality of the document can improve. Moreover, writers and translators can become aware of best writing practices for an international public.

When texts undergo translation, writing/drafting them requires an understanding of how they will be translated. Translation generally consists of

- source text analysis,
- transfer from the source to the target language, and
- target text production.

This context implies that the writer needs to be aware of how translators read and understand texts in the initial stage of the translating process. In this respect, two "planes" can be identified: a *serial* and a *structural* one (Holmes 1994, 82–83). When they operate on the serial plane, translators consider the text sentence by sentence; when they operate on the structural plane, they try to build a mind map, i.e., to find out about the organization of information in the original or source text (ST) and the links between ideas, which make a set of sentences a whole in the ST culture.

Thus, reading a text for translation involves reading with a different focus and different purposes than, say, general reading for information or pleasure. Such reading does not occur in linear fashion; rather, it is a forward and backward movement through the text as mapped by Bell (1991, 55) and based on the degree of relevance that translators give to various elements in the text. In reading and understanding a text, translators bring their linguistic, encyclopedic, and contextual knowledge to bear—that is, their *savoir* or declarative knowledge, "knowledge resulting from experience (empirical knowledge) and from more formal learning (academic knowledge)" about the ST language and culture (Council of Europe 2001, 11). When they have to process the text to come up with a translation for a given readership, they also employ their skills and know-how (*savoir-faire*)—that is, their "ability to carry out procedures" (Council of Europe 2001, 11).

Savoir and *savoir-faire* serve to bridge the gap between ST and target text (TT) language and culture. In translation, culture has come to be seen as, "[a] shared 'model of the world,' a hierarchical system of congruent and interrelated beliefs, values and strategies which can guide action and interaction, depending on cognitive context (. . .)" (Katan 1999/2004, 26). The different levels of the hierarchy are described on the basis of Hall's (1959) iceberg model of the so-called "Triad of Culture": a technical, a formal, and an informal level.

The technical level corresponds to what is visible—music, art, food and drink, dress, architecture, institutions, visible behavior (i.e., what is above the waterline in Hall's iceberg analogy). Eating and drinking habits, for example, are fairly easy to decipher when you travel to a new country. The formal and informal levels are below the waterline and more and more hidden "but also progressively closer to our unquestioned assumptions about the world and our own (cultural) identity" (Katan 2009a, 70). The formal level just below the waterline has to do with what a culture considers appropriate behavior in a given context—that is, in terms of rituals, customs, and ways or styles of discussion, art, dress, etc.

In a British context of situation, stating that something is different can imply that the speaker does not like it, but does not say so openly out of politeness. (The same is true in an Upper-Midwestern American context, to the surprise of Americans from elsewhere.) The informal level, on the other hand, determines what is suitable in a given culture depending on beliefs, values, and sense of identity. As such, it relates to orientations in action, communication, environment, time, space, power, individualism, competitiveness, structure, and thinking (Katan 2009a, 71). "Mind the gap" signs in British train stations appear to be redundant and unnecessary information to Italians, who belong to a more individualistic, high-context culture in which everyone is expected to fend for themselves—in this case, to watch their step.

The iceberg model suggests that technical culture is shared encyclopedic knowledge and can be learned and transmitted. It also indicates formal culture is rarely taught formally, and informal culture is hardly accessible in normal circumstances (Katan 2009b, 79–87). It is therefore at the level of formal and informal culture that technical communicators have to exercise care when drafting a text meant for translating.

Some examples can help technical communicators identify good writing practices in an international context. IKEA, the Swedish furniture retailer of ready-to-assemble-furniture, produces catalogues with technical information on furniture manufacture and materials used. These documents are also marketing tools that provide ideas about interior design to prompt readers to buy IKEA products. While the company is reluctant to provide information on how it creates its catalogues, research indicates that IKEA produces an outline of the yearly catalogue in standard English at the company headquarters and then "localizes" it for the different markets (Mocellin 2010–2011, 96). An interesting consequence is that the English version for the British market is different from the English version for the US market:

> POÅNG proves that style and comfort really can go hand in hand. Take a seat and you'll discover why its ergonomic design has been a firm favorite for over 30 years. But *we don't just rest on our laurels, (comfortable as they may be).*
>
> *(IKEA US 2010: 50)*

> POÅNG proves that style and comfort really can go hand in hand. Take a seat and you'll discover why its ergonomic design has been a firm favourite for over 30 years.
>
> *(IKEA UK 2010: 158)*

The US catalogue message is that tradition is good, but improvement is possible (and advisable). In its UK version, tradition is good enough not to warrant further specifications. The production of an IKEA catalogue is thus not a mere translation from English into British or American English and into other languages. Rather, it is a process of adaptation to the requirements of single markets to reflect specific values and beliefs of consumers there. In other words, the catalogue is "localized" (Valli, this volume) where localization, drawing on definitions from the Localization Industry Standards Association (LISA), "involves taking a product and making it linguistically and culturally appropriate to the target locale (country, region and language) where it will be used" (Pym 2004, 29, adapted from LISA 2003). Thus, localization is about considering local culture to provide suitable products and/or services and describing them using a language and style that can inform and convince the audience.

Localization for different markets typically involves adapting terminology where differences exist while also keeping an eye on different advertising strategies. In the following example, *nightstands* in American English are referred to as *(bedside) tables* in British English. The slogan and copy follow advertising standards in each country, with a stress on the long-held American notion that new equals improved and on the importance of personality in the US and Italian versions. The UK catalogue stresses tradition as a British national trait combined with creativity ("Traditional with a twist"):

> HEMNES—Traditional and personal
>
> Why stick with tradition when you can make it better—and brighter? Start with black or white and then give your bedroom a burst of color. The HEMNES series offers beds, nightstands, chests and a freestanding wardrobe. Mix the pieces and finishes for a bedroom that's practical but never ever dull.
>
> *(IKEA US 2010: 156)*

> HEMNES—Traditional with a twist
>
> You can love the traditional look but still keep up with the times. With HEMNES, it's easy to start with simplest white or black, and then bring on a burst of colour. Beds, tables, chests, a freestanding wardrobe—mix different pieces and finishes from this flexible range, for a bedroom that's practically traditional but never, ever dull. (IKEA UK 2010: 158)

> HEMNES—Sotto il segno della personalità
>
> Senza abbandonare lo stile tradizionale, è possibile creare soluzioni originali: parti dal nero o dal bianco, per poi dare alla camera da letto il tocco di colore che preferisci. La serie HEMNES comprende letti, comodini, cassettiere e un guardaroba freestanding: abbina i vari elementi per creare una soluzione pratica e personale. (IKEA Italia 2010: 156)

These examples show the role of technical, formal, and informal culture in the framing of a message. In each message, appropriate terminology conveys (technical) information, while knowledge of local culture dictates the wording in each case for maximum effect.

When description touches on a deeply rooted cultural value, such as family life, rewriting texts for a different market is in order. In the following example, children in the Italian catalogue play with a computer, thereby revealing the Italian fascination with technology. The focus is on the idea of a communal space. In English—and especially in the American English version—stress is on the value of the individual and individualism in the separateness of creating rooms within a room:

> **Everyone can do their own thing . . . together**
> While the kids are battling it out on the computer, you can catch up on the day's happenings. Play with textiles, lighting and rugs to create rooms within a room—and everyone gets to enjoy their own space.
>
> *(IKEA US 2010: 27)*

> **Somewhere everyone can do their own thing . . . but together**
> While the kids are battling it out on the computer, you might want to just relax quietly and catch up on the day's events. With textiles, lighting and rugs you can create different rooms within a room—and everyone gets to enjoy their own space, as well as the pleasure of each other's company.
>
> *(IKEA UK 2010: 25)*

> **Uno spazio per ognuno, da vivere insieme**
> Gioca con i tessuti, con l'illuminazione e con i tappeti per creare vari spazi all'interno di una stanza. Così, mentre i ragazzi sono indaffarati con il computer, puoi dedicarti alle attività della giornata. Una garanzia: tutta la famiglia apprezzerà.
>
> *(IKEA Italia 2010: 25)*

Further examples come from the translation of advertising. Research in this field applies Hofstede's cultural dimensions of interpersonal communication: The Scandinavian nations, the United States, and the United Kingdom are countries with weak uncertainty avoidance (i.e., the extent to which they are prepared to take risks in a cultural encounter) and low-power distance (human relationships based on equality). These translate into communication that is person-oriented, provides as much information as necessary, and stresses equal status by using a direct style (De Mooij 2004, 192–194). In the United States and the United Kingdom, advertisements exploit ambiguity and humor, and they rely on well-known personalities or easily recognizable "types" to endorse products or services. On the other hand, Italy belongs to cultures with high-power distance (human relationships based on hierarchy) and high uncertainty avoidance combined with collectivism or ways of addressing people indirectly on the basis of their status. Advertising is a form of entertainment, information is often implicit, and figurative language is used (De Mooij 2004, 194–1195).

Transcultural advertisements and commercials have general dimensions to adapt to different cultural contexts. They are developed on the basis of concepts of globalization, internationalization, and localization (Chiaro 2004, 315) or, rather, on *glocalization* as "a globally relevant ST, based on a message which will have similar impact across different cultural contexts" (Adab 2000, 224).

A final example on translation-oriented writing derives from comparing and contrasting two EU brochures with advice on writing EU documents, *How to Write Clearly* (European Commission 2015), and on writing EU documents intended for translation, *Writing for Translation* (Piehl et al. 2010). Table 4.1 summarizes the points:

TABLE 4.1 Comparison of EU advice for writing documents and writing documents for translation

How to Write Clearly	*Writing for Translation*
Hint 1: **Think before you write**	**1. Plan your text**
Hint 2: **Focus on the reader**—be direct and interesting	**2. Emphasise what is important**
Hint 3: **Get your document into shape**	**3. Show your train of thought**
Hint 4: **KISS**: Keep It Short and Simple	**4. Make sentence structure unambiguous**
Hint 5: **Make sense**—structure your sentences	**5. Avoid long sentences with a complicated structure**
Hint 6: **Cut out excess nouns**—verb forms are livelier	**6. Avoid empty verbs and 'nominalisation disease'**
Hint 7: **Be concrete, not abstract**	**7. Use the active voice wherever possible**
Hint 8: **Prefer active verbs to passive—and name the agent**	**8. Beware of noun strings**
Hint 9: **Beware of false friends, jargon and abbreviations**	**9. Use technical terms and expressions in a consistent way**
Hint 10: **Revise and check**	**10. Avoid expressions which are specific to your own culture**
	11. Use plain words
	12. Give the translator background information

Points use somewhat different language—see Hint 1, "Think before you write," and Point 2, "Plan your text," with Hint 2, "Focus on the reader," and Point 2, "Emphasise what is important." In some cases, they also follow a different order—Hint 5 in *How to Write Clearly* is Point 4 in *Writing for Translation*. Yet, in most cases, points reflect very similar, sensible advice, which largely derives from campaigns for plain language.

What strikes one most in contrast, though, are the points that one can find in one list but not in the other. Hint 10, "Revise and Check," appears to be relevant only to *How to Write Clearly*, while Hint 9, "Beware of false friends (false cognates), jargon and abbreviations," is split into two points, 9 and 11, in *Writing for Translation*. Clearly, false friends are not a major concern when documents are in one language only. In translation, however, consistency in the use of terminology and clear expression is very important. Therefore, unnecessarily complicated words are avoided. In documents intended for translation, expert technical communicators avoid culture-bound expressions. They also provide background information where necessary.

Similar points with reference to writing were made in a brochure, *Translation: Getting It Right*, written by Durban (2006) for the British Institute of Translation and Interpreting (ITI) and the American Translators Association (ATA) to guide clients as they decide how best to buy translation services:

- Conceive your documents as international ones from the start.
- Take into account the importance of style.
- Consider that in some cases pictures are much clearer than many words.
- Finalize texts before they are translated.

Advice on aspects that relate to both writing *and* translating include a number of considerations—for example, whether translation concerns a whole text, a summary, or just its gist, or whether a text is translated just for information or is meant for publishing. In the former case, translation can be quicker, while in the latter, it needs to be more accurate because it will last longer. Technical communicators need to further consider if a machine-translated version would be accurate enough to be understandable and therefore fit for purpose. They should also

- provide translators with information about the intended purpose of the text,
- show awareness that professional translators work into their native language,
- give clear indication of the language variety TT receivers speak,
- have in-depth knowledge of technical subjects,
- be aware of requirements for formatting and layout,
- show awareness of typographical differences in languages, and
- indicate a need for consistency and accuracy in terminology.

Doing so can help resulting texts better meet the related contexts in which they will be translated.

The brochure was so successful that it was translated into a number of languages (French, German, Italian, Japanese, Russian, and Spanish) and localized into American English. In this case localizers changed the wording where necessary and provided different formatting and layout using arrows and boxes to stress main points. Later another guide, *Translation: Buying a Non-Commodity* (Durban and Melby 2007, 6), followed. Here the authors stress the importance of quality standards in translating, in particular the European Translation Standard (BS EN 15038) (CEN 2006) and the American one (ASTM F 2575) (ASTM 2006), to be achieved by gaining information on the following specifications:

1. Audience
2. Purpose of the translation
3. Deadline
4. Price
5. Subject area and type of text
6. Source language and regional variation
7. Format
8. Volume
9. Target language and regional variation
10. Identification of the steps to be followed during the production phase

Again, such factors can help with aspects of quality and speed associated with a given process.

Thus, to be effective, writing for translation must be a form of "controlled" writing. In this case "controlled" is not meant in the same way as document preparation to make documents suitable for translation software. In particular, it covers machine translation, where texts undergo preproduction processing and postproduction revising, or computer-assisted translation, which is maximally efficient when language is highly repetitive and standardized.

Revising and/or Editing

This section does not supply information for editing as part of writing as it starts from the assumption that document writers have already been trained in editing. Rather, it outlines editing, revising, and reviewing with a view to providing further insight into translating, which ST authors can bear in mind when drafting documents. The underlying idea is writers can compare and contrast their practices with those of translators and eventually, perhaps, adjust them to make documents more translator-friendly.

Following Mossop (2014), editing is understood here in a narrow sense as the work of checking and changing texts, namely "copyediting, stylistic editing and certain aspects of structural and content editing" (2). By contrast, revising refers to work undertaken to check other translators' work, either as required within organizations, such as the European Union or to comply with translation quality standards, such as BS EN 15038 and its 2015 update for use worldwide, ISO 17100.[2]

In the EU, senior translators revise junior translators' work before translated documents are delivered. In terms of work after translation proper or postproduction work, BS EN 15038 and ISO 17100 divide such processes into five stages:

1. Checking
2. Revision
3. Review
4. Proofreading
5. Final verification

BS EN 15038 does not include any definition of "checking" and "final verification." "Proofreading" is the "checking of proofs before publishing" (CEN 2006, 5). "Reviewing" is to "examine a target text (2.15) for its suitability for the agreed purpose and respect for the conventions of the domain to which it belongs and recommend corrective measures" (CEN 2006, 5). Finally, "revising" is to "examine a translation for its suitability for the agreed purpose, compare the source (2.13) and target texts (2.15) and recommend corrective measures" (CEN 2006, 6).

In the relevant sections of the standard, all these stages are sequential. They start with the translator's checking her or his own work for meaning, omissions, and errors, continue with third-party revision (usually by another qualified translator) and monolingual TT reviewing[3] for suitability for purpose, and end with proofreading and final verification by the translation service provider (TSP), that is, the translation company. Mention of corrective measures in this context indicates that—unlike previous national standards such as the Italian UNI 10574 (UNI 1996)—translation quality assurance includes end-product quality. However, as is the case with all translating in general, it is difficult to envisage these stages merely as a sequence, especially as the boundaries between them appear fuzzy rather than clear-cut.

Gouadec (2007) provides a similar breakdown including checking (386), editing (387), —which he equates to revision (393)—and proofreading (391). Checks include the following:

1. Material quality checks
2. Language, style, and register quality checks
3. Technical-factual semantic quality checks
4. Transfer quality checks
5. Homogeneity and consistency checks, also known as harmonization (Gouadec 2007, 76)

Interestingly, Gouadec (2007, 70–71) envisions that the translator or the "reviser" "checks the quality of the source material" so that it is up-to-standard and is translatable. Here the professions of technical communicator and translator meet and overlap. The production of documents becomes a process in which each stage flows seamlessly into the next and boundaries blur.

According to Mossop (2014, 30–31), editing differs from rewriting and adapting in that it only requires minor changes. He lists four kinds of editing:

1. Copyediting to produce documents which comply with predetermined rules—correct for house style, spelling and typographical errors, syntax and idiom, punctuation, and usage
2. Stylistic editing to adjust vocabulary and sentence structure to the needs of the intended readership and to achieve a smooth text
3. Structural editing for the physical structure of the text—headings, lists, visuals, notes and the like. Editing for problems with the language used includes missing markers, unfulfilled announcements, empty backward references, false backward or forward references, unexplained acronyms, references to visuals, and poor paragraphing. Editing for problems with headings covers wrong headings, confusing heading system, lack of subheadings, and headings that do not match the table of contents
4. Content editing to check macro-scale content, factual errors, conceptual errors and obscure passages, logical and mathematical errors

Further checks concern consistency (i.e., to make sure that solutions, once found, are implemented throughout the text) by deciding what features need to be consistent, for example as to terminology (Mossop 2014).

Revision, by contrast, involves four groups of problems in texts:

1. Group A—Problems of meaning transfer: accuracy and completeness.
2. Group B—Problems with content: logic and facts.
3. Group C—Problems with language and style: smoothness, tailoring, sub-language, idiom, and mechanical.
4. Group D—Problems with physical presentation: layout, typography, and organization. (Mossop 2014, 134–135)

Criteria for editing and revising are similar, if not identical. One can regard them as different ways of looking at the same problems.

For an example of a translator's editing, we can turn to a column that George Monbiot (2011a) wrote for *The Guardian* newspaper. Translators then rendered it in nine target languages, including Italian, made available by the portal *VoxEurop.eu* (formerly *Presseurop*) (Monbiot 2011b, 2011c, 2011d). Monbiot's column lends itself well as an example because of its technical nature in parts and because the Italian magazine *Internazionale* published its own translation in Italian (Monbiot 2011e). Thus, juxtaposing the *VoxEurop* and *Internazionale* translations can illustrate how a text is edited and revised for translation. The German translation also confirms points noted during analysis. It is not included for considerations of space, though it is listed among the references. For the same reasons, problems of physical structure will be omitted since the fairly long stretches of text quoted should also give some idea about how the translations are presented as to layout, typography, and organization.

Monbiot's article needs to be tailored to the house style of *VoxEurop* and its concerns, i.e., issues that are relevant for debate within the European Union (see Table 4.2.). Even the original text underwent some changes to

TABLE 4.2 *The Guardian* article translated for the online portal *VoxEurop* and the Italian weekly *Internazionale*

Some greens have wildly exaggerated the dangers of radioactive pollution. [For a clearer view, look at the graphic published by xkcd.com. It shows that the average total dose from the Three Mile Island disaster for someone living within 10 miles of the plant was one 625th of the maximum yearly amount permitted for US radiation workers. This, in turn, is half of the lowest one-year dose clearly linked to an increased cancer risk, which, in its turn, is one 80th of an invariably fatal exposure. I'm not proposing complacency here. I am proposing perspective.] If other forms of energy production caused no damage, these impacts would weigh more heavily. But energy is like medicine: if there are no side-effects, the chances are that it doesn't work.	Alcuni ecologisti hanno fortemente esagerato i pericoli legati all'inquinamento radioattivo. Se altre forme di produzione energetica non causassero alcun danno, l'impatto di questi incidenti sarebbe di gran lunga maggiore. L'energia, però, assomiglia alle medicine: se non ha effetti collaterali molto probabilmente non funziona.	Alcuni ecologisti hanno enormemente esagerato i pericoli di inquinamento radioattivo. Per avere un quadro più chiaro della situazione basta leggere i dati pubblicati dal blog xkdc.com. la quantità media di radiazioni assorbite subito dopo l'incidente del 1979 a Three Mile Island, negli Stati Uniti, da chi viveva nel raggio di 15 chilometri dalla centrale è stata pari a un 625esimo del limite massimo annuo consentito per i lavoratori degli impianti atomici statunitensi. Questa soglia corrisponde alla metà della quantità di radiazioni legata a un sicuro aumento del rischio di tumori, che è a sua volta un 80esimo della dose considerata mortale. Sono dati che aiutano a mettere le cose in prospettiva. Se le altre fonti di energia non causassero danni, l'impatto del nucleare peserebbe molto di più. Ma l'energia è come una medicina: se non ha effetti collaterali, probabilmente non funziona.

meet the requirements of presentation on *The Guardian* website, instead of its print version. The headline, "Why Fukushima made me stop worrying and love nuclear power" is far from the witty, enticing headlines used for columns in British newspapers. It conforms instead to web standards, which require dull headlines making for easy literal online searches (Wheeler 2011). Moreover, paragraphs in the article are clearly spaced out to facilitate reading onscreen.

Technology and medium of transmission play an important role in deciding the physical presentation of a text. In *VoxEurop* the headline is preceded by an indication of the topic, environment, which is part of the issues debated together in the portal—with politics, society, economy, culture and ideas, and Europe and the world. The English headline, "Environment: Fukushima, The Best Ad for Atomic Power," is more allusive than *The Guardian* website original, whereas in Italian, the topic is different, as the article is listed under nuclear energy: "Energia nucleare: Come Fukushima ha reso l'atomo simpatico." *Internazionale* uses a descriptive headline, as is standard practice for the printed press.

With reference to problems of meaning transfer, the Italian *VoxEurop* translation in the example does not reflect the message of the ST (accuracy): By omitting part of the text (in square brackets in the original)—that is, lacking completeness—the argument is lost and the columnist's position appears hardly justifiable. *Internazionale*, by contrast, reports all the points in the argument.

As to problems of content, the Italian sentence that follows does not make sense as it refers to "gas impiegato per bruciare i combustibili", that is, gas—a type of fuel—used to burn fuel. Clearly, nonsense originates from a misinterpretation of *for* in "for heating fuel," which in context means "as heating fuel" and not "to heat fuel." *Internazionale* correctly renders the phrase as "il gas impiegato per il riscaldamento domestico:"

> Like others, I have called for renewable power to be used both to replace the electricity produced by fossil fuel and to expand the total supply, displacing the oil used for transport and the gas used for heating fuel.

> Come altri, anch'io ho raccomandato di utilizzare l'energia rinnovabile sia per sostituire l'elettricità prodotta dai combustibili fossili sia per aumentare l'erogazione complessiva, per sostituire il petrolio utilizzato per i trasporti e il gas impiegato per bruciare i combustibili.
>
> (*VoxEurop*)

> Come molti altri ambientalisti, ho spesso invocato l'uso delle fonti rinnovabili sia per sostituire l'elettricità prodotta dai combustibili fossili sia per ampliare l'offerta, rimpiazzando il petrolio usati nei trasporti e il gas impiegato per il riscaldamento domestico. (*Internazionale*)

A problem of content is also apparent in the following example, where a factual error is committed in indicating the date—17th century—as 15th century. However, a factual, conceptual or mathematical error clearly has much more wide-ranging consequences in different contexts, where the operating temperature of machinery, for example, can cause accidents leading to industrial injury or worse.

> Traction was intimately linked with starvation. The more land that was set aside for feeding draft animals for industry and transport, the less was available for feeding humans. It was the 17th-century equivalent of today's biofuels crisis. The same applied to heating fuel.

> La trazione animale fu strettamente collegata alla penuria di cibo: quanta più terra doveva essere adibita a sfamare gli animali da tiro nel settore industriale e dei trasporti, tanta meno ne restava per sfamare gli esseri umani. Si può dire che nel XV secolo quella fu l'equivalente dell'odierna crisi dei biocombustibili. E lo stesso si potrebbe dire dei combustibili per il riscaldamento.

As regards problems of language and style, in the *VoxEurop* translation of the paragraph, the English word order is reproduced so closely that the sentences do not run smoothly and focus is lost. Especially in the second sentence, rhetorical considerations (i.e., what would be a natural style in Italian that makes for easy reading) should lead to thematizing the phrase, "On every measure," as is done, in fact, in the *Internazionale* version:

> But the energy source to which most economies will revert if they shut down their nuclear plants is not wood, water, wind or sun, but fossil fuel. On every measure (climate change, mining impact, local pollution, industrial injury and death, even radioactive discharges) coal is 100 times worse than nuclear power.

> In ogni caso, la risorsa energetica a cui farebbe ritorno il maggior numero delle economie qualora si chiudessero le loro centrali nucleari non sarebbe il legno, né l'acqua, né il vento e neppure il sole, bensì i combustibili fossili. Il carbone è cento volte peggiore dell'energia nucleare sotto ogni punto di vista (il cambiamento del clima, l'impatto delle miniere, l'inquinamento a livello locale, i danni e le morti industriali, perfino gli scarichi radioattivi).

> Il punto è che se le centrali nucleari venissero improvvisamente chiuse non si tornerebbe a usare le fonti energetiche invocate dagli ambientalisti—la legna, l'acqua, il vento e il sole—ma i combustibili fossili. E da tutti i punti di vista (cambiamento climatico, impatto dell'estrazione, inquinamento locale, infortuni sul lavoro e morti, perfino emissioni radioattive), il carbone è cento volte più dannoso dell'energia atomica.

In the following paragraph, tailoring to the users of the translation is not maximally effective as Italian readers will have difficulty identifying, "At high latitudes like ours," as referring to both the United Kingdom and Italy. Besides, they may fail to grasp the contrast between a farmhouse in Wales and a big city, as Italians generally associate Birmingham with industry, not size.

> At high latitudes like ours, most small-scale ambient power production is a dead loss. Generating solar power in the UK involves a spectacular waste of scarce resources. It's hopelessly inefficient and poorly matched to the pattern of demand. Wind power in populated areas is largely worthless. This is partly because we have built our settlements in sheltered places; partly because turbulence caused by the buildings interferes with the airflow and chews up the mechanism. Microhydropower might work for a farmhouse in Wales, but it's not much use in Birmingham.

> Alle nostre latitudini, produrre energia pulita su piccola scala è una perdita secca. Generare energia solare nel Regno Unito comporta un madornale spreco di risorse. È un processo inefficiente e scarsamente adeguato alla domanda. L'energia eolica nelle aree densamente popolate è in buona parte inutile. Ciò è dovuto in parte al fatto che abbiamo costruito i nostri insediamenti e i nostri centri urbani in luoghi riparati, e in parte al fatto che le turbolenze create dagli edifici interferiscono con i flussi di aria. Una microcentrale idroelettrica può anche andar bene per una casa colonica in Galles, ma è assai poco utile a Birmingham.

In all of the Italian *VoxEurop* translation, but particularly in the following example, the style is not suited to an article with lots of technical references, and the terminology is at times substandard or incorrect. *Centrali eoliche a terra* is a highly infrequent term for onshore farms compared to *parchi eolici sulle coste* in the *Internazionale* version. *Centrali con serbatoi sulle montagne* for reservoirs on mountains is a rather awkward phrase with the incorrect term *serbatoi* (lit. "tanks") for the head noun. Consequently, the last sentence is difficult to understand while the *Internazionale* one makes perfect sense thanks to the standard terminology it contains (*centrali idroelettriche, alimentate da bacini artificiali*). Moreover, the mechanics of Italian do not work as *VoxEurop* fails to observe the slightly formal style which is typical of news writing in Italian—see the slightly informal terminology in *piloni, tralicci e cavi* and phrase in *mantenere accese le luci* with the more formal *connessioni alla rete, i tralicci e le linee elettriche* and *costruire nuove centrali idroelettriche* in *Internazionale*.

> Like most greens, I favour a major expansion of renewables. I can also sympathise with the complaints of their opponents. It's not just the

onshore windfarms that bother people, but also the new grid connections (pylons and power lines). As the proportion of renewable electricity on the grid rises, more pumped storage will be needed to keep the lights on. That means reservoirs on mountains: they aren't popular, either.

Al pari della stragrande maggioranza dei verdi, sono favorevole a un più vasto impiego delle fonti energetiche rinnovabili. Riesco anche a capire gli argomenti dei loro avversari: a dar fastidio alla gente non sono soltanto le centrali eoliche a terra, ma anche le nuove reti elettriche (piloni, tralicci e cavi). A mano a mano che aumenta la percentuale di energia rinnovabile che entra in rete, occorre una maggiore potenza di erogazione per mantenere accese le luci. Questo significa costruire centrali con serbatoi sulle montagne, neanche questi molto amati.

Come la maggior parte degli ambientalisti, sono favorevole ad aumentare il ricorso alle fonti rinnovabili. Ma riesco anche a capire le critiche di chi è contrario. Non sono solo i parchi eolici sulle coste a infastidire la gente, ma anche le nuove connessioni alla rete, i tralicci e le linee elettriche. Inoltre, a mano a mano che aumenterà la percentuale di elettricità prodotta da fonti pulite, bisognerà costruire nuove centrali idroelettriche, alimentate da bacini artificiali. E neanche questo piacerà alla gente.

Editing and revising translations to make them suitable for their end users is highly complicated as they may be intended for different purposes than their STs. Original texts, translations, and target-language originals can be collected to form parallel and comparable corpora to analyze using dedicated software. Comparing translations with originals and target-language originals raises possibilities to solve the kinds of problems outlined already. It also helps writers to decide on the most suitable writing styles for a given genre and topic. *VoxEurop* relies on voluntary work by translators. *Internazionale* needs to guarantee publishable quality to sell its magazine. In this latter case, editing and revising are part of the editorial policy to fit the house style of the magazine. This job can be entrusted to professionals other than translators, such as technical communicators. It is indeed quite common for newspapers and magazines to entrust "postproduction" of translation to experienced in-house journalists (Bielsa and Bassnett 2009). Editors or other internal staff members can also prepare a text—especially when it is fairly long—to make it suitable for translation for a given newspaper or magazine. Clearly, both pre- and post-editing of this kind can be outsourced to technical writers or communicators.

Revising machine translation or computer-assisted translation output is an altogether different matter. Texts intended for machine translation are both prepared for translation or "pre-edited" and only post-edited if the process is cost-effective. It is quite clear that computers still have problems at times in making sense of natural languages. Yet it is undeniable that machine translation—thanks to customization of dictionaries and work on frequency of occurrence—has made considerable progress and is here to stay. Editing computer-assisted translation requires text extraction from translation memories to work in a desktop publishing environment.

Computer-assisted translation (CAT) tools have changed the way in which professional translators work. Source and target sentence alignment has to be preserved to ensure that a translation memory is of use in the future. As a result, translators cannot operate as they do with word processing tools and keep going backwards and forwards to alter and revise parts of text as they go along. Preparing the text for translation, translating, and editing have become three separate activities, often entrusted to different professionals.

Accordingly, the concept of translation quality has also changed (Drugan 2013). Translations of varying quality can be produced starting from one and the same original text, as previous examples illustrate. As this implies that translators have to change their mindset, new professions have emerged. Technical communicators can now work as pre- or post-editors in their native language. They can seek training to write controlled language that is suitable for computer-assisted and especially machine translation. They can also specialize in editing translations produced using CAT tools or machine translation, where they will readily spot content that sounds machine- rather than human-produced. However, few corporations seem to have set up these forms of collaboration between technical communicators and translators (O'Brien, this volume).

For budget reasons, corporations often rely on machine translation without either pre- or post-editing. Any user of household appliances or computers will have come across user manuals in many languages that are hardly understandable, precisely because they are just "raw" machine translations—that is, machine translations without either pre- or post-editing or both. There is thus considerable scope for technical communicators to improve the quality of controlled language and editing and broaden their range of activities in technical communication.

Conclusion

This chapter has provided an overview of issues in writing for translation and in translators' editing and revision. The aim has been to illustrate how an understanding of the translator's profession can enhance the work of technical communicators by helping them decide how to draft documentation in an increasingly globalized world. The need for collaboration between technical communicators and translators within the framework of international communication has also emerged. Technical writing and specialist translation are not independent tasks but stages in the process of document production.

Writing, editing, and revising processes have been described from the point of view of translators' needs in a professional context and with reference to quality. Throughout, an attempt has been made to suggest how collaboration between technical communicators and translators from an early stage leads to the best quality of technical documentation in translation. Practicing such professional collaboration can start with collaborative projects during university education and training, such as through the Trans-Atlantic and Pacific Project (TAPP; See www.ndsu.edu/english/trans_atlantic_and_pacific_project/). It can continue with participation in projects for the production of technical documentation. Thus, both translators and technical communicators as professionals can improve output quality and enhance visibility of their work. Finally, this chapter has tried to show how current developments in the translation industry create new professional profiles of interest to both translators and technical communicators.

Notes

1 In 2009, the European Union (EU), as the organization with the largest in-house translation service in the world, estimated that the value of the European language industry would amount to 16.5 billion euros in 2015, up from a projected 8.4 billion euros in 2008 (Rinsche and Portera-Zanotti 2009, iii). This covered translating and interpreting (including localization and website globalization activities), language technology, subtitling and dubbing, language teaching, and multilingual support within conference organization. According to the EU, in 2012, 74% of translators were freelancers, and 60% were part-time workers, while annual salaries ranged from 6,000 to 90,000 euros (Pym et al. 2012, 3).

2 A number of quality standards exist for translation, including ISO 9001 and 9002 as applied to translation services, German DIN 2345, Austrian ÖNorm 1200 and 1201, and US ASTM F2575. These norms generally concern the workflow in companies. EU's BS EN 15038 is the first quality standard for translation to deal with individual translators rather than companies' quality. For this reason, BS EN 15038 is used here as more relevant to individual technical communicators. This standard can be cross-referenced with ISO 17100.

3 Marketing and sales staffers play a very important role in this stage of the process.

References

Adab, Beverly. 2000. "Towards a More Systematic Approach to the Translation of Advertising Texts." In *Investigating Translation*, edited by Allison Beeby, Doris Esinger, and Marisa Presas, 223–234. Amsterdam/Philadelphia: Benjamins.

ASTM. 2006. *Standard Guide for Quality Assurance in Translation*. ASTM F2575-06. www.astm.org/Standards/F2575.htm.

Bass, Scott. 2013. "Why ISO Certification (Still) Matters." *TranslationDirectory.com*. www.translationdirectory.com/articles/article1260.htm.

Bell, Roger T. 1991. *Translation and Translating: Theory and Practice*. London/New York: Longman.

Bielsa, Esperança, and Susan Bassnett. 2009. *Translation in Global News*. Abingdon/New York: Routledge.

Byrne, Jody. 2012. Scientific and Technical Translation Explained. Manchester: St. Jerome.

CEN (Comité Européen de Normalisation). 2006. *Translation Services: Service Requirements*, 1st ed. EN 15038:2006. Brussels: European Commission for Standardization.

Chiaro, Delia. 2004. "Translational and Marketing Communication: A Comparison of Print and Web Advertising of Italian Agro-Food Products." *The Translator* 10, no. 2: 313–328.

Council of Europe. 2001. "Common European Framework of Reference for Languages: Learning, Teaching and Assessment." *Council of Europe* Portal. www.coe.int/t/dg4/linguistic/Source/Framework_en.pdf.

De Mooij, Marieke. 2004. "Translating Advertising: Painting the Tip of an Iceberg." *The Translator* 10, no. 2: 179–198.

Drugan, Joanna. 2013. *Quality in Professional Translation*. London/New York: Bloomsbury.

Durban Chris. 2006. *Translation: Getting it Right*. Institute of Translation and Interpreting (ITI). www.iti.org.uk/attachments/category/54/GIR_english.pdf. American version: http://atanet.org/publications/Getting_it_right.pdf.

Durban, Chris, and Alan Melby. 2007. *Translation: Buying a Non-Commodity: How Translation Standards Can Help Buyers and Sellers*. Institute of Translation and Interpreting (ITI). www.iti.org.uk/attachments/category/54/TranslationNon-commodity.pdf.

European Commission. 2015. *How to Write Clearly*. Luxembourg: Publications Office of the European Union. https://publications.europa.eu/lastedition/725b7eb0-d92e-11e5-8fea-01aa75ed71a1.

European Parliament and Council of the European Union. 2006. Directive 2006/42/EC of the European Parliament and of the Council of 17th May 2006 on Machinery, and Amending Directive 95/16/EC (Recast). *Official Journal of the European Union*, L 157, 9.6.2006: 24–86. https://eur-lex.europa.eu/LexUriServ/LexUriServ.do?uri=OJ:L:2006:157:0024:0086:EN:PDF

Gouadec, Daniel. 2007. *Translation as a Profession*. Amsterdam/Philadelphia: John Benjamins.

Hall, Edward T. 1959. *The Silent Language*. Garden City, NJ: Doubleday.

Holmes, James S. 1994. "Describing Literary Translations: Models and Methods." In *Translated*, edited by James S. Holmes, 81–91. Amsterdam/Atlanta: Rodopi.

Huijsen, Willem-Olaf. 1998. "Controlled Language: An Introduction." Paper presented at the 2nd International Workshop on Controlled Language Applications, Language Technologies Institute, Carnegie Mellon University, Pittsburgh, PA, May 21–22.

IKEA ITALIA. 2010. *Catalogo IKEA*. www.youblisher.com/p/2151-IKEA-Catalogo-2010/

IKEA UK. 2010. *IKEA Catalogue*. http://static.onlinecatalogue.IKEA.com/2010/IKEA_catalogue/GB.

IKEA US. 2010. *IKEA Catalog*. http://onlinecatalog.IKEA-usa.com/2010/IKEA_catalog/US.

Katan, David. 1999/2004. Translating Cultures: An Introduction for Translators, Interpreters and Mediators. 2nd ed. Manchester: St. Jerome.
Katan, David. 2009a. "Culture." In *Routledge Encyclopedia of Translation Studies*, 2nd ed., edited by Mona Baker and Gabriela Saldanha, 70–73. London/New York: Routledge.
———. 2009b. "Translation as Intercultural Communication." In *The Routledge Companion to Translation Studies*, edited by Jeremy Munday, 74–92. Abingdon/Oxon, UK/New York: Routledge.
LISA (The Localization Industry Standards Association). 2003. *The Localization Industry Primer.* 2nd ed. Féchy, Switzerland: The Local Industry Standards Association. www.immagic.com/eLibrary/ARCHIVES/GENERAL/LISA/L030625P.pdf.
Mocellin, Enrico. 2010–2011. "A Linguistic Analysis of Marketing Strategies in IKEA Catalogues." Graduate thesis, Università di Padova.
Monbiot, George. 2011a. "Why Fukushima Made Me Stop Worrying and Love Nuclear Power." *The Guardian*. March 21, 2011. www.theguardian.com/commentisfree/2011/mar/21/pro-nuclear-japan-fukushima.
———. 2011b. "Environment: Fukushima, the Best Ad for Atomic Power." *VoxEurop.* March 22, 2011. www.presseurop.eu/en/content/article/564301-fukushima-best-ad-atomic-power.
———. 2011c. "Energia nucleare: Come Fukushima ha reso l'atomo simpatico," translated by Anna Bissanti. *VoxEurop.* 22 marzo. www.presseurop.eu/it/content/article/564761-come-fukushima-ha-reso-l-atomo-simpatico.
———. 2011d. "Umwelt: Fukushima, Werbung für die Atomkraft," translated by Patricia Lux-Martel. *VoxEurop.* 22 März. www.presseurop.eu/de/content/article/564971-fukushima-werbung-fuer-die-atomkraft.
———. 2011e. "In difesa del nucleare." *Internazionale*, 25 marzo: 44–45.
Mossop, Brian. 2014. *Revising and Editing for Translators.* 3rd ed. London/New York: Routledge.
Musacchio, Maria Teresa. 2006. "Quality in Published Specialized Translations." In *Insights into Specialized Translation*, edited by Maurizio Gotti and Susan Sarcevic, 173–192. Bern/New York: Peter Lang.
Piehl, Aino, Inkaliisa Vihonen, Kotimaisten Kielten, and Ehrenberg-Sundin, Barbro. 2010. *Writing for Translation.* Regeringskansliet, Translation Centre for the Bodies of the European Union. http://cdt.europa.eu/en/documentation/writing-translation.
Pym, Anthony. 2004. *The Moving Text: Localization, Translation, and Distribution.* Amsterdam/Philadelphia: John Benjamins.
Pym, Anthony, François Grin, Claudio Sfreddo, and Andy L. J. Chan. 2012. "The Status of the Translation Profession in the European Union." *Studies on Translation and Multilingualism* 7. http://ec.europa.eu/dgs/translation/publications/studies/translation profession_en.pdf.
Rinsche, Adriane, and Nadia Portera-Zanotti. 2009. Study on the Size of the Language Industry in the EU. London: The Language Technology Centre Ltd. http://ec.europa.eu/languages/news/pdf/language-industry-study_en.pdf.
Sager, Juan C. 1994. Language Engineering and Translation: Consequences of Automation. Amsterdam/Philadelphia: John Benjamins.
UNI. 1996. Definizione dei servizi e delle attività delle imprese di traduzione e interpretariato, UNI 10574. Roma: Ente Nazionale Italiano di Unificazione.

Wheeler, David R. 2011. "Google Doesn't Laugh: Saving Witty Headlines in the Age of SEO." *The Atlantic*. May 11, 2011. www.theatlantic.com/technology/archive/2011/05/google-doesnt-laugh-saving-witty-headlines-in-the-age-of-seo/238656/.

Wright, Sue Ellen. 1993. "The Inappropriateness of the Merely Correct: Stylistic Considerations in Scientific and Technical Writing." In *Scientific and Technical Translation*, edited by Sue Ellen Wright and Leland Wright, 69–86. Amsterdam/Philadelphia: John Benjamins.

PART II
Localization

AREA 3

FUNDAMENTALS, CONTEXTS, AND CONTENT

5

FUNDAMENTALS OF LOCALIZATION FOR NON-LOCALIZERS

Paola Valli

Introduction

Every day, products and services are available and sold around the globe. This does not mean the same product or service is available in all international markets (target markets) because end-users may have different expectations. The menus of the McDonald's fast food chain provide prime examples of revising products to address different local markets. McDonald's restaurants in India, for example, do not serve beef and pork products (thus respecting local Hindu values); instead, they offer the McVeggie. The McDonald's menus of other nations reflect similar local additions to the standard menu, such as the McLaks with grilled salmon in Norway, McFalafel in Israel, or the Teriyaki McBurger in Japan. Such product diversity requires geographical and cultural awareness to create products that people in different international markets can easily relate to and understand. These adaptations represent examples of localization—adapting a product line to address the expectations, preferences, and tastes of consumers in different nations and cultures.

This chapter provides readers with an overview of central factors to which individuals should pay attention when developing materials for broader, international audiences. Specifically, it discusses what it means for a company to try to sell products to different national and cultural markets. In so doing, it examines key aspects that technical communicators must consider when working on products (and related content) for audiences/users in other nations. To this end, the chapter defines what international "target markets" are and reviews the phases of the process for adapting materials for different markets. It also reviews examples of the more representative typologies of localizable products. This approach will help technical communicators understand what localization is and how it works in relation to developing effective content for users from other nations and cultures.

Overview of Localization

Before addressing the more practical aspects of content adaptation, let's examine the main theoretical concepts underlying the adaptation process, known as *localization.*

GILT

Localization is one of the four interdependent processes collectively referred to as GILT (Globalization, Internationalization, Localization, and Translation). Traditionally, the four processes are represented as concentric circles with globalization being the outermost and translation the innermost level. A different approach considers localization and translation as two interdependent but only partly overlapping processes (Scarpa 2008, 294–295). The main overlap between translation and localization resides in the linguistic transfer of the original content from the source (i.e., original) language to the target language (i.e., language into which the text is translated). The localization process involves a number of modifications on products and content at the linguistic, cultural, and technical level, and such modifications are done to account for differences in distinct markets.

For localization to be cost-effective, products should be designed in a way that allows the product and its content to be easily and readily adapted to a different market (i.e., locale) with minimum change. This strategy should be implemented from the very beginning in the product/content development cycle and continue throughout the whole production process. For example, allowing enough white space in the source documentation for possible text expansion (i.e., needing more words to convey the same concept in a different language) or including graphics with editable text can help keep localization costs within budget (Sikes 2012, 44).

Internationalization

Internationalization has been defined by the late Localization Industry Standards Association (LISA) as "the process of generalizing a product so that it can handle multiple languages and cultural conventions without the need for redesign" (Esselink 2000, 2). In essence, internationalization is the process of designing products in a way that allows them to be quickly and easily adapted to meet the specific expectations of audiences in different locales (i.e., easy to localize).

There is a fundamental distinction between internationalization and localization. Internationalization happens only once and at the stage of product design and development of materials for broader, international markets. In contrast, there can be several rounds of localization for the same product with each round focusing on adapting a product for a specific cultural market or each chosen "target market" (i.e., locale). Essentially, one takes an internationalized product (one easy to adapt) and then localizes it (adapts it) to meet the needs of a specific local market. In this way, one internationalized product can be localized in different ways and at different times to address various national and cultural markets.

Essentially, product internationalization helps prevent unforeseen cost generation and delays associated with having to revise products designed for one specific cultural market to meet the expectations of another. If a product is internationalized, it can be localized at some future time without costly retrofitting operations. Products internationalized to be used within the greater "European Union (EU) market," for example, can be more easily adapted to the marketplace of a specific nation (e.g., France) within the EU at later points in time. Such design makes it easier to develop new, culture-specific (i.e., localized) materials for specific nations as new member states enter the EU or the product becomes available in new markets where bilingual texts are needed (e.g., French and Dutch for Belgium).

When it comes to software products, three main steps can be identified within the internationalization process (Sikes 2012, 45):

1. Removal of cultural assumptions from the software design (e.g., hardcoded date format and currencies).
2. Definition of discrete layers in the software architecture such as user interface, program code and database structure. This ensures that the text to be localized (i.e., extracted from the code and sent out for translation) is well-structured and easily accessible. For example, using separate language-specific folders as well as external string tables to store localizable text makes it easy to organize and integrate translations. Separating the style of web page components from translatable elements prevents your content updates from affecting layouts.
3. Support for standards such as Unicode character encoding (such as UTF-8), so that user input and text coming from other components can be processed correctly and efficiently. This also includes choosing fonts that support user languages. For example, Arial does not support languages such as Chinese, Japanese, and some characters used in Eastern Europe.

This three-step process provides a solid base for successful localization, because it allows the application to automatically adapt to different regions without the need to rewrite the application for every language or market.

For the internationalization process to be adopted and implemented early on in the product development cycle, the company should develop a global vision in relation to international markets. This global vision is generally referred to as *globalization*, defined by LISA as addressing

> [. . .] the business issues associated with taking a product global. In the globalization of high-tech products this involves integrating localization throughout a company, after proper internationalization and product design, as well as marketing, sales, and support in the world market.
>
> *(Esselink 2000, 4)*

Thus, an important aspect of a globalization strategy is the status of the original product.

In the development phase of the globalization process, the product and its content should be treated as a "localized" instance of a locale-neutral product (Sikes 2012, 44). This means the original product should not be strictly considered as the source for all the future localized versions. While globalization is "the primary macroeconomic and upstream process driver" (Dunne 2006, 4–5), translation and product localization are the two processes that greatly influence how effectively organizations can realize the objectives of globalization.

Companies often blame localization for additional costs and delays. In truth, such factors generally result from localization tasks not being considered or addressed until the very end of a product development cycle—at a time not long before the product is scheduled to be released into international markets. This timing—in combination with the fact the original product might not have been designed with localization in mind—generally causes the actual problems at this stage. For example, it may take two to three years to develop a product in English, whereas its localization into at least 20 languages is often expected within six months of the time the parallel English product is developed (DiFranco 2006, 54).

Locale

At this point, it is useful to discuss the concept of *target market*, which represents the area where the product will be sold. This area is characterized by a collection of specific elements including language and geographical area, which are generally referred to as *locale*. Linguistic aspects (such as the language that individuals speak) alone are not sufficient to distinguish between two locales because the same language can be spoken in different countries. Consider, for example, the varieties of Spanish spoken in Mexico and Spain, which are as much or more different from each other as the varieties of English spoken in the United Kingdom and the United States. Yet, in different locations, speakers of the same language often use content/products in very different ways regardless of the language used to convey information. Consider, for example, how the laws governing data protection in Spain vs. Mexico affect how individuals in each nation can access information online. The concept of locale therefore is replete with regional, cultural, and technical aspects.

A locale is conventionally identified by double code: A linguistic label ("language code," usually lowercase) and a geographical label ("country code," usually uppercase). For example, the language code for English (*en*) can be complemented with as many as 30 different country codes representing the nations in which English is spoken (e.g., country code "CA" for "Canada"). Within this context, locale codes are not arbitrary but defined by international ISO standards. In particular, language designations generally use a two-letter code (ISO 639-1), whereas for regional designations, a two-letter, capitalized code (ISO 3166-1) is used instead. Code casing is important to avoid confusion. For example, lowercase *ca* indicates the language code for Catalan (ca_ES) whereas uppercase *CA* is the country code for Canada (en_CA).

Some examples of locale-dependent specifications are character and line orientation. While most languages use the same left-to-right writing pattern as English, languages such as Arabic and Hebrew have a mainly right-to-left directionality, and languages such as Chinese can be written vertically. Specifications also include information on numbers and dates. Indeed, numbers, dates, and currencies are examples of elements that are central to the concept of locale and designate regional and cultural conventions. (These items are discussed in depth later in this chapter.)

In sum, the concept of locale is central to the localization process because it defines the kind of translation and adaptation needed for a software or web product to be used and sold in a target locale (Esselink 2000, 1; Dunne 2006, 4). According to the definition provided by the LISA, "localization involves taking a product and making it linguistically and culturally appropriate to the target locale (country/region and language) where it will be used and sold" (Esselink 2000, 3).

From Localization Companies to Language Service Providers (LSPs)

While professional translators have always been attentive to what has become known as localization, localization as a term and distinct process emerged between the late 1980s and early 1990s. Thanks to the advent and global spread of the internet, software publishers needed to circulate their software products outside their country of origin. This created a key demand for translated and localized products. Initially, in-house translation departments and freelance translators could meet the demand alone. However, it soon became obvious that the localization of software components could not be handled as smoothly as traditional translation of plain text (Esselink 2000, 5). Text to be localized was not limited to product documentation. Rather, it also often came embedded in software source code that had to be preserved. At the same time, the size and complexity of localization projects gradually increased together with the number of languages involved. A "mixed-sourcing model" emerged (Brooks 2000, 49), where software companies would outsource the localization (i.e., translation) of all textual elements while the more technical aspects of the process were still dealt with in-house.

Within this context, localization companies (today mainly known as Language Service Providers, or LSPs) established themselves starting in the mid-1990s, and they have continued expanding over time. Some multilingual language vendors or translation companies started to specialize in software localization or other localization-specific domains and now offer dedicated services, such as engineering, testing, desktop publishing, content development, marketing services, and support services. When it comes to distributing the product, LSPs can also become specialized in areas such as localization of Adwords campaigns or Search Engine Optimization (SEO) translation.

Within an LSP, different players can be involved in a localization project. Project managers act as middlemen between the client and the translators (vendors), and they also interact with engineers, testers, desktop publishers, reviewers/proofreaders, language specialists, and any other project team member. Additionally, the size of the team can vary according to the project scope and complexity, which is usually determined by the number of languages involved. The more complex the project, the more careful the budgeting and planning need to be.

Over time, companies have kept improving their workflow and turnaround time to keep up with market demand and to address growing competition. Doing so involves continually addressing time, cost, and quality: the three main variables at play in a localization project. Unfortunately, it is often difficult to accommodate all three variables simultaneously, and the rule of thumb is "pick any two." As a result, localization efforts may vary greatly from project to project. Such factors also mean that a localization project first requires planning and proper setup.

The implementation phase of the localization process follows that of testing and updating. Ideally, the early phases of a localization project should match the initial stages in the original product development lifecycle (i.e., product concept, design and prototyping, together with the drafting of the documentation) and continue during the alpha and beta testing and further product developments. At minimum, the localization team should be *involved* in those product development phases in order to raise awareness about the future linguistic pitfalls. However, this is often not the case because localization is generally moved down to the very last stages of the product lifecycle when most work has already been completed. Additionally, the way in which companies include localization in their product and content development lifecycle depends on their level of localization maturity—a reference stage (out of five possible stages) based on the company's behaviors, processes, and activities (DePalma 2007).

Simship

Corporate awareness about GILT and adequate strategic planning can help companies rationalize their development lifecycle to support simultaneous shipment of products into all target locales. In the past, it was not unusual to have a lag of eight (sometimes 12) weeks between a core release (for the US market) and an international release. Thanks to the adoption of globalization strategies, the lag could progressively be reduced and eliminated. Such developments often allow for products to be shipped simultaneously to all international markets. This process of simultaneous shipment, or *simship*, refers to the practice of releasing a new product worldwide at the same time, instead of releasing the original version in its home market and then proceed to localize and distribute it (Asnes 2010, 22).

If simship is the goal for the company, additional planning is required not just on the developer side but also on the localization side. The translation of the software product and related documentation has to start while the core or original release is still being developed (Esselink 2006, 26). This is one of the main challenges for the company, as localization is often treated as a tail-end process, where it is generally associated with delays and higher costs. For the average employee, it is also much easier and faster to think only in terms of one's own locale when developing a product and its related content. However, experience shows that including localization earlier on in the product development process helps uncover critical problems, which can be costlier to repair at a later stage. Simship goes hand in hand with internationalization because eventually the success and implementation of simship is dependent on the quality of the internationalization efforts and practices.

Simship is not a must for every product, and its use depends on the company, the company's maturity, and its global outreach. A company that is relatively new to global markets will likely emphasize business agreements as opposed to simship. Conversely, experienced and global companies are more likely to engage in achieving simultaneous shipment of their products, for example, when it comes to releasing security updates or critical patches (Asnes 2010, 23) or simply just releasing a new technology product such as a smartphone.

Connection to Professional Practice

In the preceding sections, product localization has been discussed in terms of strategic planning and product life cycle. Awareness of the project scope, however, is also vital if internationalization and localization efforts are to be included in the content life cycle. In today's world, consumers are information-hungry, but they are also exposed to information overload. Product-related content has come to cover an extremely broad spectrum of domains. Yet actual product content (e.g., software strings, user interfaces, and running text) is complemented by traditional product documentation including user manuals, quick start guides and online help. When it comes to developing the product, designing the interface, and writing product documentation, communication writing skills can be required of engineers, scientists and other professionals—and not just professional and technical communicators.

Interestingly, the majority of today's products comes with significant marketing content, the most obvious example being a product page or a website. Such marketing content can range from blogs, advertisements, and social media presence to rich media (i.e., those with which users can interact as opposed to solely animation) and possibly a strategy for search engine optimization (SEO). Marketing strategies are deeply affected by localization. A survey conducted by Common Sense Advisory, for example, indicates 75% of consumers surveyed are more likely to buy a product with content in their own language: content they understand and can relate to (DePalma, Hegde, and Stewart 2014).

When end users speak the same language as the developers and technical communicators, there are no communication issues, as long as the original contents are properly drafted. In this scenario, technical and professional communicators typically pay attention to clarity, consistent terminology, and information load. Yet distributing new products to international markets often means introducing new terminology, concepts, and communication style into the target language, and this implies that local languages adapt to the influx of English (Hellstern, Gasova, and Safar 2013, 36). For example, in the mid-1990s when the internet started to become available to the public, verbs like "*surfing/browsing the internet*" were introduced and later translated into other languages like "*navigare in internet*" in Italian. In more recent times, the advent of Twitter ensured that words like "*tweet*" were adopted in many languages and new verbs—like *tweettare* in Italian—were added to everyday vocabulary. This phenomenon brings with it the need to at least know about the existence of the technology that the terms refer to.

If organizations wish to market a product or service in another country, they need to ensure that the target audience believes that the product was originally designed and developed for them. The product does not just consist of technology. The terminology associated with a product also plays a major role for consumers, so much so that the language of the application may be as important as its actual functionality (Hellstern et al. 2013, 37). Additionally, terminological changes introduced in the original English version may not always be successfully transposed to the target language versions. This is because there is not necessarily a direct correspondence between a word in one language and a word in another. In fact, there rarely is.

Language, moreover, is just one of the aspects that contribute to the overall feel of the product. Numbers, graphics, and colors are other elements that contribute to a product's feeling familiar and its integration into the target market. The following sections examine such elements for products that are prime candidates for localization: software applications, computer games, and web content.

Software Localization

Three components of software frequently undergo localization: software or program files, online help, and product documentation (Esselink 2000, 11). Each component requires a different approach to localization due to the related level of technicality.

The analysis of source material usually comes in a localization kit that includes a running version of the English product, a build environment for the software and online help (including source files), and a documentation folder. If available, earlier releases of the software can also be provided together with glossaries and translation memories (Esselink 2000, 19). Glossaries and translation memories are resources that help translators produce a localized version of the product that is consistent and in line with the original contents.

Translation memories (TM) in particular are collections of original and translated text portions (segments) that are fed into a special software, allowing translators to reuse previously translated words or sentences. Table 5.1, for example, presents an example of how elements in a TM system can serve as the foundation for creating a troubleshooting checklist.

In this example, the first (left-most) column in Table 5.1 contains two similar but not identical sentences; the second (middle) column contains two completely different statements, whereas the third (right) column contains exactly the same wording.

With translation memories, the translator will not have to translate the text of the third column twice because the software will recognize that the sentences are identical. As for the first column, the software will recognize a certain degree of similarity, and it will offer the first translation as a partial translation suggestion for the second (fuzzy match). The second column, on the other hand, will have to be translated from scratch in both cases.

The reusability principle also applies to the original English version. This means that, when drafting a text, reusing the same sentence several times can

TABLE 5.1 Examples of text to be used with translation memories

"Camera not found"	Camera not connected correctly	Check USB cable and connect it properly to the device and PC! Start the software again!
"Device not found"	Device was not recognized	Check USB cable and connect it properly to the device and PC! Start the software again!

help ensure that the translated version is also consistent. To increase consistency through repetitions, technical communicators can rely on the Simplified Technical English (STE) specifications and use restricted grammar rules and a controlled-language dictionary. This approach means that content can be reduced by up to 30%; translation costs, up to 40% per language. For a user manual of about 300 pages (approximately 54,000 words), the related savings can range from US$1,980 for one language to US$23,760 for 12 languages (Braster 2009, 51).

Some general rules for producing a well-prepared source document can help save time and money during the translation phase. Content should be kept as flexible as possible, and authors should bear in mind that for each source sentence a one-to-many relationship will arise (i.e., between the original sentence and each of its target language translations) (Argondizzo 2011, 46). In addition, formatting issues can considerably slow the translation process, particularly when translation memory systems are employed. Some examples of formatting issues include manual page breaks, hard line breaks within sentences, and soft line breaks in lists as they can affect the quality of the text segment to be translated. Similarly, adding colored markings or internal comments in the text and using nonstandard or inconsistent date/time/number formatting can slow down the process because requests for clarification may arise. Building on these points, Keller (2011, 43–45) recommends some best practices for maximizing sentence reuse with translation memories. These include avoiding using PDF files as source document format for translation and replacing multiple blank spaces with tabs. They also encompass using automatic hyphenation and adopting uniform spelling of specialized terminology and abbreviations, which is very important for consistency in the target text.

Software Application

Software application usually involves the translation of user interface (UI) components, such as dialog boxes, menus, and strings (Esselink 2000, 58). Localizing these items may not seem particularly challenging as they generally contain short messages or individual words; however, the very lack of context while translating software strings is a high-risk source of mistranslations. Therefore, always be sure to add descriptive comments about commands or polysemous words. Additional challenges emerge due to space limitations and the phenomenon of text expansion (or reduction) between languages.

Text expansion entails how the length of a word or overall expression in one language can increase in size or overall verbiage when the same concept is presented in another language. For example, a simple one-word command, such as *Quit*, will more than double in length when translated into German (*Abbrechen*) (Dietz 2006, 125). Generally speaking, German words and phrases tend to be about three times longer than their English equivalents, while

Chinese characters take up less space than English words (Wittner and Goldschmidt 2007, 10). Translating from English into the Dutch, Finnish, German, or Indic languages in particular involves an expansion in the volume of the text of up to 40% (DiFranco 2006, 48). This can have consequences for dialog boxes and other user interface items, such as buttons, that have a fixed size.

During the translation of localizable content, specific translation tools and software can be used to resize some interface items and parse source files in order to separate localizable from non-localizable content (DiFranco 2006, 49). Where interface resizing is not possible, abbreviations can be used (e.g., using [SELEZ. STAMP.] instead of saying *selezionare stampante* for *select printer*). However, abbreviations have been known to make the text or functionality more obscure to the target reader. Abbreviations and acronyms are often used in English to make the text more concise, but they may not translate well in other languages and are likely to pose problems (Wittner and Goldschmidt 2007, 10).

As authors tend to work only with the original source document (i.e., the original version of the text), they tend to expect other versions of the manual to be produced in a straightforward manner. However, certain language-dependent phenomena need to be taken into account, such as differences in a language's alphabet. For example, some languages have additional letters, if not a completely different alphabet altogether. Consider, for example, the letters "æ," "ø," and "å" in Danish and Norwegian or the umlaut vowels in German (ä, ö, ü) or special consonants in Polish (ć, ń) and Turkish (ç, ğ). This factor can impact not only the way characters are visualized (character encoding) but also the way in which information is presented (sorting order). And the results can be problematic.

The chosen encoding scheme for one language may not support a number of additional characters in another language (Wittner and Goldschmidt 2007, 10). In addition, special characters are handled differently from one language to another. In Polish, extended characters are inserted after their non-extended version of the character (i.e., . . . n, ń, o, ó, p, . . .), while in Swedish they are all placed at the end of the alphabet (i.e., . . . x, y, z, å, ä, ö). In Hungarian, characters may be written with one, two, or three characters (i.e., . . . a, á, b, c, cs, d, dz, dzs, e, . . .), and this matters for sorting at the level of the graphical user interface (GUI) and documentation (Wittner and Goldschmidt 2007, 10).

There are instances where specific linguistic awareness is necessary to prevent future problems in the localized product. Shortcuts and placeholders both relate to the source code of a program. Keyboard shortcuts, for example, help users speed up some operations while placeholders promote the reuse of strings. Placeholders (or variables) are combinations of characters in a string that are replaced by a given word. Usually, a variable replaces a number, a product or company name, dates, and version numbers (Argondizzo 2011, 46), but in fact, any text can be replaced by variables.

The string "Wait until the light turns %d" is an example of a placeholder where "%d" stands for red, yellow, or green. Placeholders of this kind work very well with languages like English with a simple morphology (i.e., where word endings do not change much from one sentence to the other). For other languages, however, one has to take into account gender and number agreement and case endings, which makes the use of such variables much more complicated (Wittner and Goldschmidt 2007, 10). Awareness of problems in the rendition of placeholders in other languages may save time and money if the number and nature of placeholders are well considered in advance. Other potential issues emerge if field codes are used instead of normal text for items such as dates. In order for the day and month to be translated, the field needs to be removed; otherwise, the text will be automatically displayed in the system language.

Shortcuts for specific programs are chosen so that they can be meaningful for the user, for example, by using a combination of keys with the initial letter of the command to be executed (e.g., ctrl + B to make a font into bold). When switching to another language, the original letter may no longer be meaningful to the user (e.g., *bold* is translated as *grassetto* in Italian and as *fett* in German). The first option for addressing this situation is to change the shortcut to match the target language (e.g., ctrl + G or ctrl + shift + F, respectively). Alternatively, the original shortcut (ctrl + B) can be left untouched to promote standardization across different localized versions of the program.

Possible linguistic issues have been the main focus so far, but additional challenges to a localization project can also come from cultural elements. From a software developer's perspective, cultural elements can be linked (e.g., numbers and dates). These numeric issues include currency symbols that should be placed before or after the amount to be converted in the appropriate currency. By the same token, measurements, be they the imperial, US, or metric system, should be provided in the appropriate unit for the reader. Decimal separators also change. In the US system, the decimal separator is a dot (full stop/period) while a comma is used to separate thousands. In most of Europe (excluding English-speaking countries and Switzerland), it works the other way around: a comma marks the decimal spot while a dot (full stop/period) separates numerals at the thousands spot.

The label *date* is used here to comprise all references to time. Differences can emerge in the way a locale measures time (e.g., AM/PM or 24-hour format) or counts the day of a week. In the United States, Sunday is considered the first day of the week, whereas according to the international standard, a week starts on Monday and ends on Sunday. Similarly, date formats can vary from country to country (e.g., mm/dd/yyyy in the United States, dd/mm/yyyy in many European countries, yyyy/mm/dd in Sweden). If ignored, such differences can lead to problems such as missing a deadline or appointment or misinterpreting a sell-by date, as the date 04/11/2013 can be read as both April 11th or November 4th, depending on the locale.

After all localizable material has been translated, the testing of the translated version of the software takes place and can include linguistic testing, functionality testing, and compatibility testing (Esselink 2000, 21). The linguistic testing is still considered part of the translation process and aims at verifying the content of dialog boxes and menu items. The functionality testing carried out by engineers is very similar to the testing that was done on the original product. This testing focuses on verifying the compatibility of the localized version with the target operating systems, application, and hardware standards. Finally, compatibility testing ensures that the product can work with other products available in the target language.

Online Help

Online help usually represents the largest and possibly most demanding component in a localization project. It contains an overview of the product, procedures, dialog box descriptions, reference material, and all information explaining how the product works (Esselink 2000, 166; Monti 2007, 182). Hypertext versions of online help have replaced traditional printed documentation, which means that cross references are used to point the users to related sections by citing them. They are typically used inside sentences and can range in length from single words to whole sentences, sometimes involving multiple variables (Argondizzo 2011, 47).

In order for online help to be useful, consistency should be maintained between the software application and the references in the online help. This strategy involves using consistent terminology across all product components. This approach can be challenging at times because similar content may be drafted by different units/departments within the company and take different forms (e.g., online help, documentation, web pages, training material, and so forth).

Because of its volume and the need for consistency, online help lends itself to be used in conjunction with translation memories. These tools, in turn, are a valuable aid in localization projects with highly repetitive contents. Terminological work can also be a valuable resource to prepare a multilingual product-specific glossary. Such glossaries, however, should be validated by the client. These glossaries list recurring terms in the interface and/or documentation, reference material for terminology, and updated terms.

Potentially challenging elements are graphics and visuals, because they are as important as the written word. They can be used both in the online help and product documentation. Several types of graphics may appear, such as generic graphics, screen captures (with or without translatable text), and illustrations (Esselink 2000, 347). The original graphics are not necessarily applicable to all localized versions of the product because of the localized interface and/or locale-dependent hardware components. Computer keyboards exemplify the latter.

In contrast to a standard US keyboard layout (QWERTY), a German keyboard swaps the Y-key with the Z-key (QWERTZ), while a French keyboard swaps the A-key with the Q-key (AZERTY). Other letters differ in position as well. Additional changes and swaps can be found in the punctuation and special character keys. Awareness of different keyboard layouts may prevent keyboard input from not being recognized.

Product Documentation

Product documentation can range greatly in type and size, depending on the product. It can include an installation manual, an online manual, a getting-started guide or quick-start guide, and collateral material (Esselink 2000, 289). Product documentation originally came in printed form, but it now comes mainly in electronic format with the printed material limited to a leaflet or quick-start guide (Monti 2007, 183). One of the challenges when writing product documentation is conciseness, particularly when it comes to web writing. A shorter text means reducing the word count and therefore localization costs. When translation budgets do not allow for translation of the whole text, only the most relevant parts should be identified and localized (Sichel 2009, 3). In general, translating the whole English text might not always be necessary for the target audience. Careful consideration of which contents need to be localized can keep translation within a reasonable budget.

Localizing printed material also includes keeping track of the standard paper format in the target locale. Doing so can require some layout redesign (e.g., from standard 8 ½ -inch by 11-inch US letter to the 210-mm by 297-mm A4 page size). A desktop publishing (DTP) team is often involved in the localization of documentation and is responsible for fixing the layout of translated material, editing graphics, converting, or generating output in multiple formats and performing quality checks (Esselink 2000, 323).

Text and screenshots are still present, but additional graphics and texts are used as marketing material. In this respect, colors and images may require careful handling. For example, in China and Japan, white is the color of death and mourning, while in the Western countries it is associated with festive events such as weddings. Images of parts of the body, such as eyes, hands, or feet, can be perceived as offensive in some countries (Wittner and Goldschmidt 2007, 10). Even geography can become an issue if maps are used, because sensitive geopolitical issues may be brought up if a given representation of borders and geographical areas or a geographic name is chosen over another (e.g., Persian Gulf vs. Arab Gulf).

Authoring environments and content management systems (CMS) give technical communicators the possibility of using different tools, which promote content reuse also for product documentation. These tools are particularly helpful when it comes to producing documentation for similar products or different releases of the same product, in which case conditional formatting may be helpful. This approach is used, for example, when the same product comes in different versions with increasingly more features. Product warnings, caution statements, and some of the user instructions would be the same across all models. Conditional formatting allows properly keeping together the features of each product version to be used in the same block of text, thus avoiding the risk of mismatch (Argondizzo 2011, 48).

Game Localization

Game localization is a special area for localization and one where simship often plays a key role. The primary reason behind the marketing of computer games worldwide is the need to generate profits to cover the high cost of game development. Large teams of programmers, designers, graphic artists, and other experts are involved in projects that can last several years. Because of fierce competition and computer games' limited shelf life, game developers come under enormous pressure (Dietz 2006, 125). The game industry relies heavily on localized versions of games. In fact, such localized versions generally make up about 50% of the global revenue of the industry, though for individual companies the percentage can range from 25% to 70% (Edwards 2011, 18).

Simship implies parallel development of the game in its original as well as localized versions. Achieving this objective means working on the game interface, error messages, video or audio, dubbing, subtitling, information files, maps, help files, tutorials, gameplay hints, playguides, peripherals setup files, keyboard layout guidelines, read-me files, patch files, and so on (Dietz 2006, 121). Yet, lack of cultural awareness can lead to problems when marketing a product. Such a situation can be particularly acute when products are not duly localized and do not account for game content that could be considered offensive or inappropriate or even unlawful in certain international contexts. This situation entails not only removing potentially problematic content but also introducing more familiar and local aspects so that the target audience can better identify with the game. For example, using a baseball home plate as an icon for the home screen may not be immediately obvious for end-users from countries where baseball is not as popular as in the United States (Dietz 2006, 128). In contrast, linguistic understanding has diminished as a major obstacle in computer games, because English is increasingly accepted among gamer crowds.

Web Localization

In terms of volume, web localization has now surpassed software localization and is sometimes referred to as e-localization, content localization, or web content localization (Jiménez Crespo 2008, 6–7). In the early days, web texts were static and changed little over time or according to the user group. Today, there is no stable and final version of a text; thus, different users may be presented with a different text when accessing a website. Web pages tend to be increasingly dynamic and are often enriched with software functionalities and applications (Jiménez Crespo 2008, 63). Within this context, content is much harder to separate from software, which in turn is tied to a specific platform or application, while the role of the user interface has become particularly prominent (Filip 2012b, 41). A new type of localization project has emerged that focuses on programming and publishing.

This situation means new opportunities, such as parallel publishing of multilingual content, and fully automated localization processes, thanks to the adoption of content management systems (CMS). At the same time, it also implies that requirements will be very different from those of traditional software localization. This is largely due to shorter turnaround time and translation of content from other languages into English (i.e., opposite translation directionality) (Esselink 2006, 27; Monti 2007, 179; Jiménez Crespo 2008, 30).

One option to deal with parallel content development involves the use of a "pivot" or "relay" language for translation (i.e., a language that is used as a bridge for translating into other languages) (Filip 2012b, 42). Traditionally, all localized versions were produced starting from the (English) core release, but in fact there are language pairs that are closer to one another than to English. Localization efforts will likely be less costly in this case, and language technologies can be used more efficiently as the number of languages involved increases. For example, among the languages belonging to the Slavic group—Belarusian, Bulgarian, Czech, Macedonian, Polish, Serbo-Croatian, Slovak, Slovenian, and Ukrainian—some similarities among the languages can be exploited (Filip 2012b, 43). For example, some Ukrainian users may find it easier to read instructions in Russian or Polish, rather than in English, not to mention that translation rates for these combinations are still very competitive.

Web content is very dynamic and is by no means limited to technical texts. Web localization also entails the translation of legal texts, advertisements, and subtitles, which makes localization closer to multimedia translation (Jiménez Crespo 2008, 29). Given the wide spectrum of possible web contents, websites require different localization strategies. Due to the e-commerce phenomenon, websites need to be kept continually up-to-date with the latest products and services and localized for the target markets. Updates affecting dozens or hundreds of files occur in real time and need to be attractive to a diverse and localized audience. In addition to content and linguistic elements (e.g., colloquial language, verb usage, personal pronouns), graphical elements such as logo, banner, pictures, image position, graphics, and animation also play an important role (Jiménez Crespo 2008, 63).

Competencies of localizers are no longer linked just to computers and languages but also cultural and political considerations, in that culture plays a much greater role in content localization. Cultural awareness in web localization also includes adapting the payment methods to those preferred in a given locale. Aside from established systems common to all target markets, some payment methods may be more popular in one locale and not so much in another. For example, in Germany many people still prefer to wire money, so users will expect to be given this option (Wittner and Goldschmidt 2007, 10).

Numbers were previously mentioned in terms of measurements, currencies, and dates, but they also play a role in addresses and telephone numbers. The positions of house numbers and zip codes may vary, just as a telephone number can be broken down in different ways according to country code, area code, and subscriber number. This situation is relevant in particular if the user is required to fill in an online form. Name conventions are also to be considered as the order of name and surname may change from country to country (Wittner and Goldschmidt 2007, 10). In Hungary, for example, as in China, family names normally precede what in English we call a person's first name (the so-called eastern name order). In most Western countries, this is not the case. (If the name Christopher Columbus were to be translated, it would become Kolumbusz Kristóf in a Hungarian context.) In the United States, it is common for people to have a middle name but is much less the case in Europe.

A whole website may eventually have too much content for it to be localized fully. Analyzing web traffic will highlight which pages are reached via search engines, usually driving most of the traffic. When it comes to making decisions about localization, understanding how your users consume content and services may be a key piece of information to have.

Projections for the Future

When the localization industry established itself, companies still published content in a linear form and, most importantly, were fully in control of their message and brand. Localization has become more than language and culture. Global customer satisfaction and business requirements have gained importance and additional activities such as support, usability adaptation, and community moderation are also required (Beninatto 2013). The industry focus is shifting towards user and customer experience with customers increasingly in control. They do not just expect information to come in their language; in addition, they can choose the company they prefer among several options. They also often expect real-time support. Moreover, they share their user experience with the world. As a consequence, content complexity also increases with interactive documents and advanced visualizations. Content creation and localization need to speed up, which means that workflows need to be improved and updated for content development as well.

Content strategy is aimed to ensure that the right content is created at the right time and for the right device. Localization requirements, however, are often still lost along the way or completely left out. The new standard is continuous localization or on-demand translation (Beninatto 2013) to which volunteer translation should be added. Users and enthusiasts are willing to embark on the translation of content and interfaces to allow fellow countrymen and women to use the product in their native language including minority languages. Such approaches have been notably adopted by social platforms such as Facebook, LinkedIn, and Twitter. In this new scenario, where individuals are directly involved in the product life cycle, neither simship nor the traditional bulk localization model, with its unique combination of price, time, and quality, is actually viable. However, simship still works for big software companies that continue to develop their product in one core language (Filip 2012b, 39–41).

As a consequence of consumers taking charge, communication has become more informal. Additionally, style is changing in all languages, growing to resemble the writing style in English (especially American English), which is renowned as more "relaxed and personal" than most of the Central and Eastern European languages (Hellstern et al. 2013, 38). (For example, in these languages, the formal version of the pronoun *you* is still used, and idioms and jokes sometimes need to be toned down.) New products are increasingly presented in a friendlier tone to create a closer link with marketing and end users, even though manuals are not expected to go through major rewrites (Beninatto 2013).

There is as well a major shift happening in the style, especially of tone and terminology, used in recent software. The language used for software products is constantly evolving with both gradual and more radical changes. These originate in the source language and are then transposed to the localized products, which means that localized language always lags a little bit behind (Hellstern et al. 2013, 36). Professional publishers (or localizers) tend to have a different attitude toward the original English text than a volunteer translator has. The latter will tend to be more faithful to the original production, whereas the professional will produce a text that has lost its indigenous flavor (Filip 2012a, 51). Examples that are culturally unintelligible are deleted or replaced with more familiar ones, and parts of the text that are regionally irrelevant are also removed.

An interesting phenomenon for localization has been the spread of mobile and, in particular, touch-screen devices. These products have partly shifted the traditional paradigm in localization. First, these devices have spread across all age groups, thanks to their intuitiveness. The main communicative element is not so much the user interface language but rather the visuals (i.e., icons and images). In this scenario, the importance of menus and text is scaled down because the iconic power of touch-screen devices reduces the overall learning curve and could help break the language barrier.

In addition to content and visuals, touch-screen devices work thanks to some coded gestures to be performed on the capacitive sensitive device. Such gestures have become globally understandable. However, in some countries or locales, some level of customization might still be necessary, for example, in the case of bidirectional languages (e.g., Arabic, Persian/Farsi, or Hebrew). In these languages, text is written from right to left while numbers or text in western scripts are written from left to right (Musale 2013, 35). If the text and interface in a traditional localization project had to be adapted to a right-to-left reading pattern, direction-based gestures and instructions for touch-screen devices might also change.

New contents and apps promoting content sharing among users suggest a further increase in the use of touch-screen devices in the future. Localization will affect the new applications and also the actual content shared among users through social media, games, and news. Sharing content has nowadays become extremely easy and quick. As a consequence, turnaround time for translation and localization has been further reduced, and any new digital content is bound to be short(er)-lived (Musale 2013, 34).

With such diverse scenarios, some level of standardization, e.g., for screen size, usability, accessibility, and security will be necessary. Standardizing message formats using core standards can help manage the multilingual information flow and, in particular, create platforms to handle multilingual content. Multilingual content is now produced on a daily basis, and it should be possible to keep track of each piece of content whether original or translated (Filip 2012a, 55).

As one might expect, such factors mean that standardization becomes necessary for more advanced forms of text and content chunking (e.g., lower than file-level) and reassembling. Content splitting (i.e., dividing up content and related tasks among different individuals) among several outsourcers is often necessary, particularly if volunteer translation is involved (Filip 2012b, 39). Standards help not only device manufacturers and application developers but also the localization industry. Different operating systems, disk space, and available infrastructures must be taken into account (Musale 2013, 36). For example, connectivity through WiFi is very common in some geographical areas, while in developing countries other connectivity methods might be required.

Conclusion

Traditionally, localization was associated with software programs, computer games, and websites, but recent advances in technology have brought about significant changes, placing customer satisfaction in a more prominent position and advancing a whole new range of mobile and touch screen devices. As a result of such factors, localizers need to constantly update their skills and competencies to keep up with these trends. Yet, for them to produce high-quality, localized products, source material needs to be drafted appropriately. This chapter aimed to provide an overview of what localization is and how it works to make technical communicators aware of their role in the content development cycle. Ideally, such knowledge can allow for better content strategy to be put into place.

Localizers, including professional and technical communicators, generally work in larger teams. Attention to stylistic and terminological consistency is paramount to allow localizers to do a good job and often is one of the most common issues that localizers face while working. Style guides can be a very useful tool to ensure a coherent and cohesive text. A style guide for localization needs to take into account the specific end-user, company and country standards, as well as geographic and cultural suitability.

In addition, a style guide for technical communicators can be a useful starting point to reduce localization challenges at later stages and allow localization professionals to concentrate mainly on the locale-specific aspects. Style guides need to be targeted to the specific product or product category. They can also cover topics such as digital texts and style or the localization process involving many languages. Style guides offer generic recommendations. A distinction, however, is needed between "content" and "interface" text, the former being different on each page and used to communicate directly with the end user. Interface text, on the other hand, provides the framework for the actual content in the form of buttons, banners, pop-up messages, and so forth.

General guidelines are consistent in promoting the basic requirements in scientific and technical writing: clarity, conciseness, precision, and objectivity. These features ensure brief, concise, informal, clear, accessible, and specific digital texts that take into account semantic-lexical and stylistic aspects, syntax, and typography, as well as iconic-visual aspects (Duque 2000, 21–22, in Jiménez Crespo 2010, 2).

References

Argondizzo, Peter. 2011. "Managing Variable Text in Translation." *Multilingual* 22, no. 6: 46–48.

Asnes, Adam. 2010. "The Business Why and How of Simship." *Multilingual* 21, no. 5: 22–23.

Beninatto, Roberto S. 2013. "Localization Trends for 2013. No News, Good News?" *Moravia*. January 15, 2013. http://info.moravia.com/blog/bid/259158/Localization-trends-for-2013-No-news-good-news

Braster, Berry. 2009. "Controlled Language in Technical Writing." *Multilingual* 20, no. 1: 50–52.

Brooks, David. 2000. "What Price Globalization? Managing Costs at Microsoft." In *Translating into Success*, edited by Robert C. Sprung, 43–57. Amsterdam/Philadelphia: John Benjamins.

DePalma, Donald A. 2007. "Moving Beyond the *ad hoc*racy of Localization." *Multilingual—Getting Started Guide*, October/November, 6–8.

DePalma, Donald A., Vijayalaxmi Hegde, and Robert G. Stewart. 2014. *Can't Read, Won't Buy: How Translation Affects the Web Customer Experience and E-Commerce Growth*. Cambridge: CSA.

Dietz, Frank. 2006. "Issues in Localizing Computer Games." In *Perspectives on Localization*, edited by Keiran J. Dunne, 121–134. Amsterdam/Philadelphia: John Benjamins.

DiFranco, Carla. 2006. "Localization Cost." In *Perspectives on Localization*, edited by Keiran J. Dunne, 47–66. Amsterdam/Philadelphia: John Benjamins.

Dunne, Keiran J. 2006. "A Copernican Revolution." In *Perspectives on Localization*, edited by Keiran J. Dunne, 1–11. Amsterdam/Philadelphia: John Benjamins.

Duque, García M. 2000. *Manual de estilo. El arte de escribir in inglés cientifico-técnico*. Madrid: Paraninfo.

Edwards, Kate. 2011. "Levels of Game Culturalization." *Multilingual* 22, no. 6: 18–19.

Esselink, Bert. 2000. *A Practical Guide to Software Localization*. Amsterdam/Philadelphia: John Benjamins.

———. 2006. "The Evolution of Localization." In *Translation Technology and Its Teaching*, edited by Anthony Pym, Alexander Perekrestenko, and Bram Starink, 21–29. Tarragona: ISG, Universitat Rovira i Virgili.

Filip, David. 2012a. "Localization for the Long Tail: Part 1." *Multilingual* 23, no. 7: 51–55.

———. 2012b. "Localization for the Long Tail: Part 2." *Multilingual* 23, no. 8: 39–44.

Hellstern, Alfred, Katerina Gasova, and Libor Safar. 2013. "Bringing a New Tone to CEE Product Localization." *Multilingual* 24, no. 2: 36–39.

Jiménez Crespo, Miguel A. 2008. "El proceso de localización web: estudio contrastivo de un corpus comparable del género 'sitio web corporativo'." Ph.D. dissertation, Universidad de Granada.

———. 2010. "Localization and Writing for a New Medium: A Review of Digital Style Guides." *Revista Tradumàtica—Traducció i Tecnologies de la Informació i la Comunicació* 8: 1–9.

Keller, Nicole. 2011. "Creating Translation-Oriented Source Documents." *Multilingual* 22, no. 7: 43–45.

Monti, Johanna. 2007. "Localizzazione: il ruolo e i saperi della traduzione." In *I saperi del tradurre. Analogie, affinità, confronti*, edited by Clara Montella and Giancarlo Marchesini, 173–197. Milano: FrancoAngeli.

Musale, Shailendra. 2013. "Touchscreen Devices in the Localization Industry." *Multilingual* 24, no. 1: 33–36.

Scarpa, Federica. 2008. *La traduzione specializzata. Un approccio didattico professionale*. 2nd ed. Milano: Hoepli.

Sichel, Barb. 2009. "Planning and Writing for Translation." *Multilingual—Getting Started Guide*, July/August, 3–4.

Sikes, Richard. 2012. "Localization: The Global Pyramid Capstone." *Multilingual* 23, no. 7: 43–49.

Wittner, Janaina, and Daniel Goldschmidt. 2007. "Technical Challenges and Localization Tools." *Multilingual—Getting Started Guide*, October/November, 10–14.

6

HOW STREETSMART MARKETERS TRY PHISHING CUSTOMERS

(Mis)Communication or Missed Communication through Loan Words

Birthe Mousten[1]

Introduction

> Pierre: "Mademoiselle Isabelle has such a nice-looking face."
>
> Jean: "Yes, it is nice enough; it is so Danish though, c'est une visage a la Danois, a la Vimmelskaft par Di[2]; however, I think I could get used to her face, if only she did not speak Danish."
>
> (*Holberg n.d.: Act.3, Sc.1*)[3]

Translating words into another language can be difficult. Even more, the non-translation of "loan words," i.e., a word taken directly from another language (*Oxford Learner's Dictionary*) or borrowed but only partly adapted for another culture (*Merriam-Webster Dictionary*), can cause misunderstanding. The use of such loan words also carries implications about the relationships between languages in terms of superiority and inferiority. Here, non-translated items might seem "superior," and their import means they need to appear in the original language vs. need to be translated. The preceding quotation reveals that this is not a new phenomenon: French had a superior status relative to Danish in the 17th and 18th centuries, when loan words and loan phrases were used to create distance between the learned (i.e., those who knew French) and the peasants (i.e., those who did not). In the quotation, we even see the young man express contempt towards the young girl because she speaks Danish.

Hierarchies between languages are just as prevalent today as they were four centuries ago—perhaps more so due to the speed and reach with which online media now allow cultures to interact. For technical communicators, such linguistic aspects have a range of implications. The first is how loan words affect the translation process and what they mean for how technical communicators should create content that is "translation ready." Second, loan words can affect how technical communicators use English to convey information and access ideas in international interactions. Communicators in both situations can benefit from an understanding of loan words, how they come to be, and how they are used. This chapter provides an introduction to a method for, as well as actual research on, the use of loan words and, in turn, provides the basis for the communication needed to recognize and, ideally, address various aspects of loan words when people work in various global contexts.

Language, Loan Words, and the Context of This Study

To examine the topic of loan words, I will consider the role that English generally plays in other languages. Specifically, I will review the adoption of English loan words in domains across genres, such as the specialized language used by, for instance, banks and stock exchanges. In my case, it is commercial enterprises.

The Role of English in Other Languages

English started out as a major recipient of words from other languages, but the pendulum shifted to English being a language donor to other languages. This shift was spurred partly by the massive colonization that created the British Empire and partly through the rise of the United States as a world power in the past 100 years. To understand this shift, Filipovic (1996) claims that the receiving languages "borrow English loans, adapt them as Anglicisms and subsequently integrate them into their vocabulary" (39). The process of incorporating English into other languages occurs in different ways, including hegemony and minority, or as a sign of status in the recipient language.

The tussle between hegemony versus minority issues is probably most important in countries where English and one or more local languages exist in a society, e.g., a former colony. In such cases, the hegemony of English might have been pressed on local cultures, thereby placing English in a position of status in comparison to the local language(s). In such cases, English could be seen as the language of oppression and hegemony, whereas the local languages could be linked with resistance, conflict, or ethnicity, or a combination. In this context, English and the local language(s) might also show a stark power asymmetry, with English speakers enjoying certain benefits that non-English speakers do not (Eriksen 1992, 313–316).

People who know English and can express themselves in English seem to have a higher status than people who cannot. India can serve as an example here, for English, since the days of the British Raj, has traditionally been the language of those in power and those who have landed an education and a good job. Another present-day example is when young subculture groups (e.g., gamers) adopt certain English phrases that then slowly spread to the mainstream language (Lønsmann 2009). Some of those phrases in Danish would be "min computer lagger" ("my computer is lagging"), or acronyms, such as "LOL" (for the English "laughing out loud"). It may also be a generation marker when young people use far more swear words than older people (Rathje 2012). Heidemann Andersen (2002) explains this as a generation gap when she says that it is a wish to be young with the young.

In other cases, English loan words appear in subcultures where the language is used to create identity and group solidarity akin to what clothing and music have done previously (Preisler 2003). The wish to be among the trendsetters could also explain the use of certain English words and expressions in Denmark today. The Danish headline "Modeshow på museum er *sooooo last season*" [in English: "Fashion show at themuseum is sooooo last season"] (Dalsgaard 2015) represents such usage, where "sooooo last season" employs from English a recent American/British slang phrase, which describes something that is no longer "in style." This example also represents English used as code switching (i.e., mixing languages, in this case English and a local language)—something common in certain business groups for reasons of power or elite identification (Fernández 1990) or as a class marker (Matus-Mendoza 2002).

English Loan Words in Domains

English in language domains has been debated for two decades (Ejsing 2009, Davidsen-Nielsen et al. 1999, Jarvad 1999, Svane 2010), and the related loss of local language has produced the concept of "domain loss" (Haberland 2005, Laurén et al. 2002, Rasmussen 2008). The idea works as follows: Loan words travel in language domains shared only by domain specialists (e.g., electrical engineers), who might use loan words instead of local words (e.g., use the English word "circuit" instead of the Danish "kredsløb"); business domains (e.g., finance or shipping) or different technical domains (e.g., nanotechnology), where a lack of local terms often tends to lead to loan word patterns in Danish communication. Another example might be when stock traders talk about "EBITDA" (i.e., the English expression "earnings before interest taxes depreciation and amortization") in Danish. Not only is it used as an acronym, it also covers an abbreviation that is not related to the Danish language.

Such situations can affect how nonspecialists in other language groups participate in certain processes connected to a particular domain. Consider the example of a stock trader and the average person both wanting to invest money in their pension funds. The language gap becomes a knowledge gap and creates a divide between those who know and can act, because they know and can use the correct loan words, and those who do not know and have to depend on others, namely those who have such knowledge to act for them. Therefore, domain loss can create two levels of conflict. The first involves the use of English loan words and the lack of, or suppression of, Danish equivalents in society. The second encompasses differences between expert communities and the public at large. In these ways, domain loss is not only an issue between languages, it also creates a divide between expert communities and laypeople (Selsøe-Sørensen 2004).

To examine such factors and the transfer of loan words in more detail, I decided to study the use of English-language loan words in non-native English contexts. Specifically, I sought to address two primary research questions:

- RQ1: Do non-native speakers of English define and use English loan words in the same way that native English speakers do when communicating in commercial contexts?
- RQ 2: If non-native speakers use English differently in commercial contexts, what happens when they define and use loan words?

By answering such questions, technical communicators working in English can begin to understand if loan words can be used at all, in moderation, or should be avoided altogether in a given context. This knowledge, in turn, can guide approaches to everything from creating global content in English to producing English-language source texts to facilitate translation. Such factors could lead to new or revised content creation and content management approaches that help organizations more quickly and effectively share information with different international audiences.

Research Approach

In conducting this research, my goal was to determine how well English loan words are adopted by Danish speakers in Denmark. To do so, I selected examples of words from a weekly, unsolicited, household- and mail-distributed, multi-page brochure covering daily-necessity purchases and some seasonal and leisure activities. These public mail distributions occur weekly, and the brochure was aimed at nonexperts (i.e., laypeople), but it often used domain-specific terminology—and related English loan words—to provide descriptions of products and services. The particular brochure was from the chain megamarket BILKA, with a supermarket, a garden center, and some specialty shops under the same roof. The domains covered in the brochure included everything from TV sets and sports clothing to food.

From the brochure, I filtered out the loan words (the English words that appeared in the Danish brochure) to try to effectively represent each domain-related product category. The particular loan words were English in spelling, meaning, and context, with the exceptions noted in Table 6.1—cases where an inflection was needed. Words such as *compatible* rendered in its Danish version *kompatibel* were not included, because I considered them partly or wholly "naturalized."

TABLE 6.1 Word checks in Danish dictionary

Word	*Constantly updated online (dsn.dk)* Retskrivningsordbogen *(Dansk Sprognævn 2013)*
Boardshorts	N/A[1]
Sneakers	N/A[1] . . . *(rubber shoes1)*
Streetsmart	N/A
Coleslaw	[1)] *(a cabbage salad) in contexts with coleslaw, for instance coleslaw dressing*
Phishing	N/A[1] . . . *(attempt at luring confidential or personal information by means of email or the like)*
Frosted	N/A[1)] / *(with a matted surface; frosted glass* [frostet glas][2]
Full suspension	*Suspension (suspending; mix of liquid and fixed particles), in context with suspensions, e.g. declaration of suspension*[3]

1) N/A = Not available, or not available at first. Later entries stated in my translation in italics.
2) When I first started my work, this word was N/A, but it now appears under the inclined form *frostet* when searching *frosted.*
3) This word was listed under suspension, but with a different meaning from another domain.

Having identified possible loan words, I then checked the official Danish source, *Dictionary of Correct Danish* [Retskrivningsordbogen] online (www.dsn.dk) to determine if a particular loan word was listed (Dansk Sprognaevn, n.d.). This gave me some discretion as to whether my words should be included or rejected as loan words, because if I, as a native speaker, saw the word as a loan word and it was not listed at dsn.dk, it would most likely be a loan word. Through this process, I identified 24 problematic loan words or terms, seven of which I examine in this chapter (see Table 6.1). I then used this information to develop a questionnaire that I distributed to respondents in order to assess these loan words in relation to the project's two research questions.

Selecting Respondents

The questionnaire was distributed to five classes, a total of 122 students, in Danish high schools and colleges (i.e., students of language, business, and math-science), and 96 respondents returned the questionnaires legibly filled in. Specifically, the five classes comprised three high-school classes with language and math-science students and two college classes with corporate subjects as their main focus. In Danish high schools at the time of testing, the students had a minimum of six years of training in English in the Danish undivided elementary and secondary school system, while some students had a year or two more. This was largely the case for the college students too. This baseline reflected their level of proficiency in English as an L2 subject. Choosing students across high school and college in this way covered the age group and educational span to avoid a language or math-science or business bias in the population of respondents.

It is often asserted that young people who are exposed from early childhood to English through computer programs, movies and TV series, or English as a school subject adopt and use English as a second language (L2 language) seamlessly in their communication with native English speakers (L1) and other L2 English users. The age range of respondents was 17–24 years of age with only a few students being older. This older group was still below 30 and constituted only about five percent of the respondents.

The questionnaire was distributed to students located in Aarhus, the second largest city in Denmark (population approximately 273,000), and it included schools and colleges located in the Jutland region of Aarhus with some students from the surrounding small towns and countryside. Statistically, this might give a small variation in the results. Taking the homogeneity of Denmark into account, however, I argue this variation must be limited, because Denmark is a small country with 5.5 million people, and the distance between large towns and the countryside is very small compared to other places in the world.

In reporting results, my statistical work aims at showing trends only. Linguistics has a tradition of a high extent of qualitative analyses with few respondents. Marketing analysis has a tradition of either a high extent of quantitative analyses with hundreds of respondents or qualitative analyses with fewer respondents, depending on the purpose. My objective here is to show trends and variety in language development and use (i.e., the qualitative approach), but to avoid bias and wrong conclusions, I aimed at getting a fairly high number of respondents. My aim was thus not to make rules or set preconditions but to show how trends and variation influence the use of loan words in advertising.

Questionnaires

Data collection for this project involved two questionnaires. To supervise how students completed the questionnaire, I either visited the participating classrooms or instructed teachers how to perform this process, which went as follows:

- Questionnaire 1, used to assess the student's immediate understanding of the loan word, was distributed first. When it was completed and returned, Questionnaire 2, used to assess domain and context knowledge of the term, was distributed.
- Respondents were given approximately ten minutes to fill in each questionnaire (i.e., 20–25 minutes total to complete both).
- Students were not allowed to access dictionaries or the internet, nor could they talk with fellow students, when completing the questionnaire.

The overall purpose of both questionnaires was to determine the respondents' understanding of the loan word: at first their own immediate understanding, if any, and then the real-context-associated understanding. Table 6.2 shows how

TABLE 6.2 Method of inquiry and evaluation of results

	Questionnaire 1: Own understanding of the word						*Questionnaire 2: Real context*	
List of words	**Method**	***Intuition*** Results	**Method**	***Definition*** Results	***Own context*** Results	**List of words in context**	**Method**	***Definition*** Results
	Word Association:	Some None ____ 96	**Word definition and context**	Correct Some idea No answer Wrong ____ 96	Correct Some idea No answer Wrong ____ 96		**Domain verification or domain change**	Correct Some idea No answer Wrong ____ 96

Black text: questionnaire results; gray bold text: method of inquiry, cf. figs 4 and 5. The graduation in toning shows the connected methods and results.

both questionnaires were designed and notes the purpose (kinds of data) each was designed to collect.

Questionnaire 1 began with a free word association (labeled intuition) to determine if the students had some association with the word, or none. For example, students were asked to write down their first association with the word "sneaker." Some wrote "shoe," others "basketball," and still others "chocolate." The first two associations reveal that the word is recognized immediately in its correct context (i.e., as athletic footwear), whereas the third shows that there is doubt, either about how to read the word or about the meaning of the word.[4]

For marketers, it is important to obtain the best possible understanding of the communication coupled with the balance between *recognition* and *novelty*. The novelty effect may justify the use of loan words. In the end, however, the message must be understood for the marketing communication to be successful. Thus, the recognition rate at *intuition* level (i.e., when people immediately see the word and create a meaning) may be lower than the target of 90% set for the *real context* (i.e., when people have a full context to base their assumptions on).

The question for *intuition* in the questionnaire prompted an association, "Which other Danish or English word do you connect with this word?" and then the heading *Intuition*. The questions for *definition* and *context*, as shown in Figure 6.1, prompt the respondent to create a suitable definition and a self-made context. As such, these definitions can be compared with word association methods where "the first response may act as a stimulus for the second response thus eliciting a chain of responses" (Cremer et al. 2011, 192).

The question for *definition* was, therefore, "What do you think the word means?" and then the byline with the word *Definition*. The purpose of this column was to make the respondent establish as precise a definition of the word as possible. In contrast to the *intuition* column, this *definition* column could be compared to existing dictionary sources to see the overlap with existing Danish sources.

The question for *own context* (i.e., how would the respondents use the word themselves) was, "In which connection do you think the word is used?"

The questionnaire originally in Danish, here translated (except loan words)

Word	Which other Danish or English word do you connect with this word?	What do you think the word means?	In which connection do you think the word is used?
	Intuition	Definition	Context
Boardshorts			
Sneakers			

FIGURE 6.1 Questionnaire 1: Own understanding of the word

TABLE 6.3 Questionnaire 2: Real context

Word	*Real context of the word*	*What do you think the word means in this context?*
Boardshorts	Island Haze *boardshorts* for men	
Sneakers	*Sneakers and* basket boots	
Streetsmart	Shoes for the *streetsmart*	
Coleslaw	Deli BBQ chicken or 2 pcs. *Coleslaw*	
Phishing	Norton protects your PC from *phishing*	
Frosted	Cupboard with *frosted* glass lid	
Full suspension	Size 24 or 26" MTB with *full suspension*	

The loan words appear in italics, the translated words from Danish in plain style

followed by the heading *Context*. The respondent was then prompted to use the word in a self-made context, so that a typical meaning of the word would be expressed. *Definition* and *own context* should thus be seen as continued elaborations of the respondent's intuitive understanding of the word. Thus, it sometimes revealed from which domain the respondent knew the word.

Table 6.3 shows Questionnaire 2, which was handed out separately so that no back-correction was possible.[5] The word in Questionnaire 2 was shown in the actual advertising context in which it appeared. Thus, the respondents had to define the word from the *real context*. If a change occurred in the responses compared to Questionnaire 1, I could establish that a correction or enlargement of the existing mental lexicon took place. Although the real context is, perhaps, very short, as in "Size 24" or 26" MTB with *full suspension*," this is the extent of communication to which the customers are exposed to the term (except graphics) and from which context they are expected to understand it.

From a marketing perspective, for the communication of the advertisement to be successful, my target for *understanding* (*correct* and *some idea* combined, see Table 6.3) in Questionnaire 2 with the *real context* is set at 90%. The 90% target even leaves room for novelty to accommodate the first movers in a market but also ensures that understanding is attained by the broader range of customers.

Overall Evaluation of Terms

To show how I computed the results, I use the example of *boardshorts*, which appears in Table 6.3. In this example entry, there is thus the same flow in the table as in the questionnaires. Additionally, the jump from the first to the second questionnaire has been marked by a sharply indicated vertical line (see Table 6.4).

TABLE 6.4 Evaluation methods coupled with results for *boardshorts*

	Questionnaire 1: *Own understanding of the word*						*Questionnaire 2:* *Real context*	
List of words	**Method**	***Intuition*** Results	**Method**	***Definition*** Results	***Own context*** Results	**List of words in context**	**Method**	***Definition*** Results
	Word Association:	Some 46 None 50 96	**Word definition and context**	Correct 34 Some idea 6 No answer 53 Wrong 3 96	Correct 33 Some idea 6 No answer 54 Wrong 3 96		**Domain verification or domain change**	Correct 47 Some idea 28 No answer 19 Wrong 2 96

For the sake of brevity, I have shortened the questionnaire questions into short headings.

For *boardshorts* (trunks worn by surfers) the recognition rate under *intuition* was 46 of 96 respondents (= 48%). Some of the doubts and misreadings under *intuition* in Questionnaire 1 proved this: *a shot that you can drink*; *something to do with IT*; *a smartboard*. Such misunderstandings vanished when the real context was introduced (e.g., when readers encountered a context of clothing). Under *real context*, the *correct* and *some idea* columns included 75 respondents, corresponding to 78%. It should be noted here that "boardshorts" is an example of influence, not just from British English or American English, but from Australian English (*Oxford English Dictionary*), where the surf communities emerged. The word has travelled to other surfer communities around the world, as well as into the fashion business.

For ease of understanding, I have converted all absolute numbers to relative figures (percentages), and I have introduced diagrams to visualize the results. Figure 6.2 shows in diagrams the results from questionnaires 1 and 2, except the *intuition* method. The reason why *intuition* is not measured on a level-of-understanding scale is that intuition can never be deemed right or wrong, compared to definitions and contexts, which have to follow commonly understood conventions and contexts. *Intuition* can give an idea of the respondent's organization of the mental lexicon in relation to the word in question.

The diagram should be read from left to right to see the development in the respondents' level of understanding. Note that the *definition* and *own context* columns show data from Questionnaire 1 and *real context* shows data from Questionnaire 2. Throughout the diagram, there is a distinction between *understood* and *not understood*. This distinction is made, on the one hand, between the two leftmost bars (*correct* and *some idea*), which are deemed to be *understood*. On the other hand, the two rightmost bars (*no answer* and *wrong*) are deemed to be *not understood*. However, even inside the *understood* and *not understood* categories, some of the diagrams show shifts.

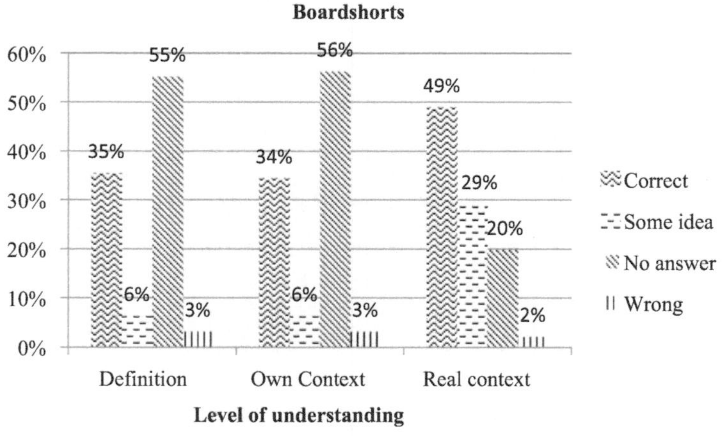

FIGURE 6.2 Results for *boardshorts*

The *boardshorts* example shows that, although many might recognize the word, or part of it, at an early stage, they depend on a *real context* to define the word with more certainty. In the *understood* category, the *definition* figures were (35% + 6%) = 41%, and the *own context category* was (34% + 6%) = 40%, respectively. As a whole—even with improvement in understanding—the (49% +29%) = 78% *understood* category under *real context* is not satisfactory compared to a marketing target of 90%.

Results and Analysis for Selected Terms

In this section, the remaining six terms used for this study are described according to the same method.

Results and Analysis for Sneakers

From the start of the project, the term *sneakers* proved to be very well understood and incorporated in the language by the age group. The progression confirmed this trend (see Figure 6.3). The recognition rate was 71% (not displayed in Figure 6.3, because intuition cannot be deemed right or wrong). An intuition might just as well point towards a different domain, in which the word might be right. In such case, it tells us something about the primary reference of the word, which might not correspond with the actual context situation), and the jump between the understanding levels from about 90% in *definition* and *own context* to 98% in *real context* did not show any significant change. The word must be deemed fully accepted in the Danish language, given the 90% target of understanding in a *real context*.

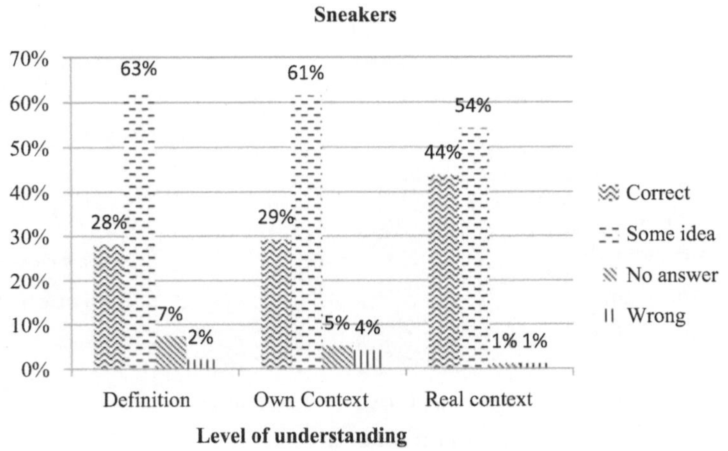

FIGURE 6.3 Results for *sneakers*

However, there was a parallel understanding of this term for a small group of the respondents, who presented a dual answer. In this context, the intuition box proved its worth, because it gave the clue to the wrong definition. Some of the respondents mixed up the word *sneakers* with the word *Snickers*, a chocolate bar. This result can be ascribed to phonological and morphological resemblance. The *Snickers* word was advertised heavily on Danish TV with the slogan *Snup en Snickers* [pick a Snickers] some years ago, and this probably triggered the association to chocolate, which was weeded out after seeing the word in real context.

Results and Analysis for "Streetsmart"

The recognition rate was only 56%. Many respondents started writing the Danish direct translation of the word in the *intuition* section: *gadesmart*. This immediate and literal translation was probably what sprang to mind naturally as the best way of explaining the word. In some cases, the respondents probably guessed at the meaning via a direct translation. This might be a problem because, according to the Danish online dictionary, the naturalized loan word *smart* in Danish basically means (in translation) either 1) good-looking fashionable clothes, and therefore attracting attention; or 2) good thinking—works effectively; or 3) a little bit too clever and fast (often derogative); or 4) elegant and intriguing, but only on the surface. The main emphasis in Danish is clearly on the first definition: wearing smart clothing (Den Danske Ordbog n.d.).

Logically, *streetsmart* in Danish would mean wearing good-looking, modern, fashionable, hip clothing, which aligns with the definition and context given by the respondents and also the context used by the Danish marketing agency. A search for *streetsmart* via Google Advanced for Danish texts yielded only some fuzzy results for Danish but with the main idea being linked to clothing and looks. Figure 6.4 shows the results for *streetsmart*. There is some uncertainty in

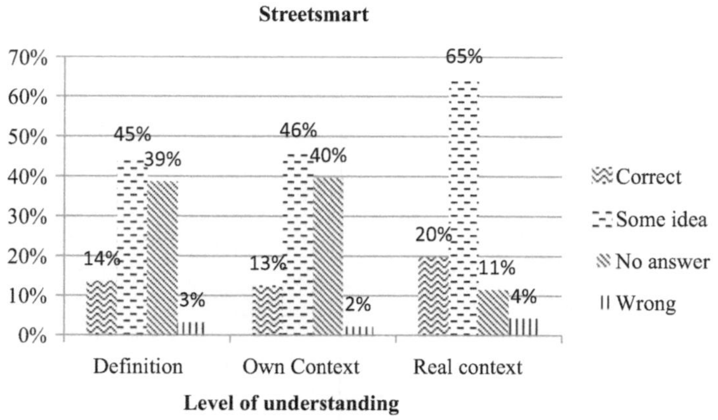

FIGURE 6.4 Results for *streetsmart*

responding to the word in *own context*, but as soon as the *real context* is shown, the results show a jump (the incremental development of especially the second bar in the diagram).

Checking *streetsmart* in English is troublesome. Merriam-Webster's online dictionary shows only the collocation *a streetsmart teenager* and then refers the reader to the synonym *streetwise*. A check of *streetwise* in the dictionary produces the definition and context *having the knowledge needed to survive in difficult and dangerous places or situations in a city*. The collocation given is *a smart, streetwise cop*.

Collins English Dictionary has the entry *street smart* with the synonym *streetwise*, too. It is defined as an informal American-English word with this definition: "Someone who is street smart knows how to deal with difficult or dangerous situations, especially in big cities." The example given is, "He is street smart and is not afraid of this neighborhood." Google Advanced searches agree, but the contexts often involve some kind of *streetsmart technology*, where technology can be anything from IT equipment to tires.

When we compare the English and the Danish meanings of *streetsmart*, the understanding in Danish is much narrower (basically linked to clothing) than in English (used primarily in connection with technology and being able to cope). The meanings in Danish and English are congruent on an ad-verbatim basis, but the most frequent use of the meaning differs. Only a few dictionaries comment on the frequency of the words in the language in question. Therefore, we might call these terms semi-false friends.[6]

A false friend (i.e., a cognate with a different meaning) is a word that looks the same in, for instance, English and another language and which usually had the same meaning, but which has taken on different meanings or a different frequency over time. For example, in English the word "editor" means someone who reviews and revises texts, while the cognate French term "éditeur" means someone who publishes texts: that is, a publisher. Semi-false friends are common in languages, especially today when so many have taken in English loan words and then used them in a narrower or skewed meaning. In a Danish context, the expression *to be street* was triggered by the term *streetsmart*. When this meaning was presented, it was as a synonym of the American primary use of *streetsmart*.

To sum up, the word *streetsmart* functions as a loan word in Danish because the word is English in appearance. The understanding in the Danish mental lexicon is congruent with the Danish marketers' use of the word, albeit not fully congruent with the English mental lexicon's preferred use of the word. The rate of *understanding* rises from the *definition* and *own context* stage of 59% to 85% in *real context*.

Results and Analysis for Coleslaw

For the term *coleslaw*, the recognition rate was 51%. The *understood* categories revealed some guesswork on the part of survey-takers. This became visible

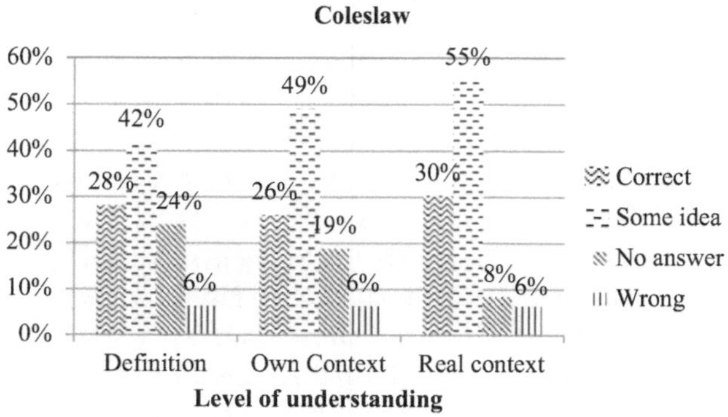

FIGURE 6.5 Results for *coleslaw*

from the general *own context* examples: for instance, "a kind of food," "a salad." While these guesses were correct, there were also wrong guesses, such as "a vegetable," "Russian salad," "meat," "chicken," "polar bear," "theory," and "Cole's law," even as other respondents mastered writing the whole recipe. I had some problems in evaluating the results for *coleslaw*, because the answer "some kind of salad" would be listed under *some idea*, whereas "Russian salad" would be wrong, because it exists as a different type of salad. However, it is possible that some of the respondents thought of "Russian salad" as falling in the category of *some kind of salad*. In general, however, I have always taken the positive view from the respondents' questionnaire answers. As shown in Figure 6.5, the rate of *understanding* from *definition* and *own context* rose from 70% to 85% in *real context*. Again, the *real context* results did not quite match the 90% target and are thus not quite satisfactory.

Results and Analysis for Phishing

The results for the word *phishing* were markedly different from the results for some of the other words. The recognition rate was 38%, which is unsatisfactorily low, and the dominant right-skewed appearance for the *not understood* bars in the *definition* and *own* context columns in Figure 6.6 shows the conversion of lack of understanding to actual understanding in *real context*. Questionnaire 1 showed that almost nobody could get even close to what the word meant, or they left the response sections blank. The responses ranged from "to fish," "a fisherman," "a hole in one's ear," to something that smacks of the domain, "something with a PC," "a virus," etc.

Most of the wrong answers related *phishing* to real fishing without relating the word to anything else. In this case, we see an example of phonological resemblance, which blurs the distinction between the two terms *phishing* and

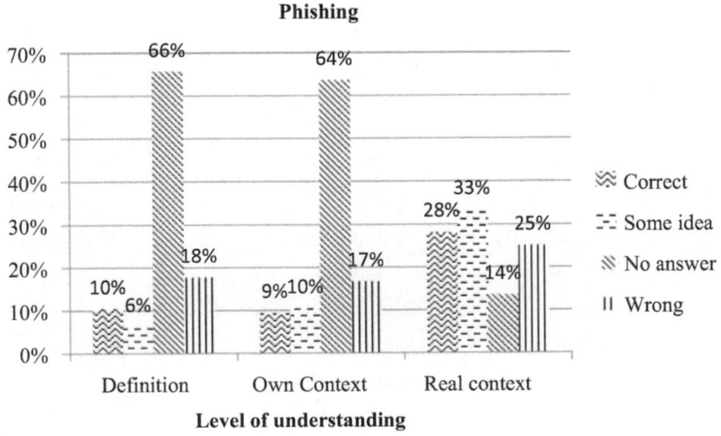

FIGURE 6.6 Results for *phishing*

fishing. It needs to be investigated to which extent the same mistake would occur among native speakers of English. However, *phishing* was also connected to the word *piercing.* Whether this was a phonological or morphological problem cannot be determined from the questionnaires.

One might argue that the answers here are a bit tricky in that the word *phishing* actually has been created from the word *fishing* but is used in a metaphorical sense of fishing out somebody's codes and passwords from their computers. This might show how much the context of a word helps people to understand and define a loan word. The actual development from the *definition* and *own context* categories rose from 16% to 61% in *real context.* Compared to the 90% marketing target, this is still not an acceptable situation.

Results and Analysis for Frosted

The understanding of the term *frosted* was blurred, ranging from "frozen," "Kellogg's frosties," "doughnuts," to "a computer screen that freezes." (As Figure 6.7 shows, the recognition rate was only 46%.) However, most individuals gave up on the word until they saw it in context; yet, even then, the picture is blurred. Again, the morphology of the word and lack of a deeper understanding of the word stem and its different inflections lure the respondents into diverse relations to the mental lexicon. "Frost" is an ancient word in Danish. Being a cognate of the same word in English, it is a true friend in most instances. For example, when a branch is white because it is frozen, a Dane would say that there is "frost" on the branches. In Danish, we would also talk about "frost in the air," and we would talk about "frostvarer" (literally "frostwares") for food kept in the freezer. However, the transfer of these meanings

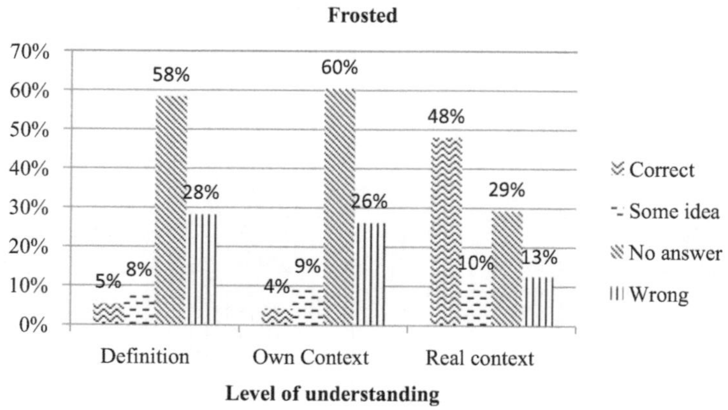

FIGURE 6.7 Results for *frosted*

to the semi-metaphorical meaning seen in the phrase "a frosted glass" has only recently occurred in Danish, in contrast to English. The level of *understanding* from *definition* over *own context* to *real context* again showed an increase from as little as 13% to 58%, which is considerably lower than the 90% target. The understanding of the loan word in the advertisement is therefore not satisfactory.

Results and Analysis for Full Suspension

For the term *full suspension*, the recognition rate of 43% is very low, and for this term many of the columns in Questionnaire 1 were left empty or contained incorrect responses. This example again indicates meanings registered differently in the English and Danish languages. In Danish, the word is linked exclusively to the meaning of being laid off from your position temporarily. This actually puzzled some respondents, because the word *full* did not make sense in this connection. The respondents' intuitive answers included such words as "banned from the job," "sent home," "fully equipped," "all included insurance," "right to return," "money back," as well as the words "gearing" and "full gearing."

The *real context* gave some of the respondents an idea of the meaning of the word but also convinced a majority that their first understanding was wrong, without giving them the right context. Thus, the development throughout the questionnaires was surprising. This was the only example in which the development throughout the questionnaires was negative, meaning that the respondents were more puzzled when they had filled in both questionnaires than they were in the rest of the cases when they gained a deeper insight and understanding. (See Figure 6.8.)

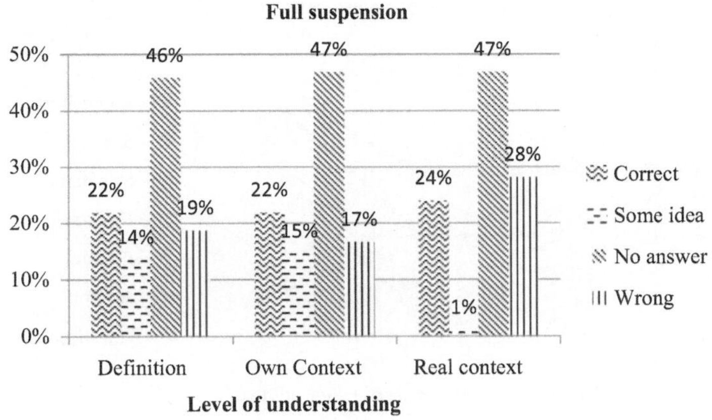

FIGURE 6.8 Results for *full suspension*

The development for *full suspension* shows a decrease in understanding, in contrast to the other words. The *definition* and *own* context figures show a level of *understanding* of about 36%, but this understanding is linked to another language domain. However, the 25%, who with *real context* did get an idea of the meaning of the word, managed to expand their mental lexicon to comprise the word, but that still leaves 75% with a wrong guess or without any clue. Compared to the desired 90% *real context* target, the 25% *understood* result is totally unacceptable.

Overall Evaluation of Terms

On the basis of the seven terms, the overall trends between the *definition, own context*, and *real context* stages may be evaluated.

Figure 6.9 shows the *understood* categories (*correct* and *some idea*), from *definition* over *own context* to *real context*. The different loan words and their recognition and understanding percentages are put into the figure to see which words are well-known loanwords and which words are virtually unknown. The words that are best known beforehand in the Danish language show only a small increase in the level of understanding (e.g., *sneakers* and *coleslaw*) whereas the words that are less integrated in the Danish language from the *definition* and *own context* stages jump to a much higher stage when put into a *real context*. This is the case for the words *boardshorts, streetsmart, phishing* and *frosted*. For all the words, except *full suspension*, there is an increase in *understanding* from the *definition* stage to the *real context* stage. *Full suspension* is a total failure in terms of finding a place in the mental lexicon, and the real context is of no help. From the marketing target of 90% in *real context*, only the word *sneakers* is acceptable.

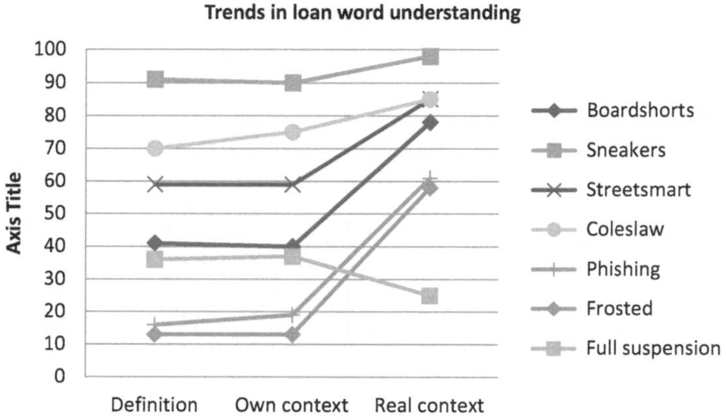

FIGURE 6.9 Overall evaluation of the loan words

Perspectives for Content Creators and Translators

To put these ideas into perspective, I might suggest that the chapter's title, "How Streetsmart Marketers try Phishing Customers," would most likely be understood by speakers of American English as "how streetwise marketers try to hack into customers' computers to get information." A Danish interpretation might be "how fashion-clad marketers try to find new customers." Admittedly, I have not taken into account that many individuals would probably read the title with some kind of irony, but I hope the example suggests the effects of using loan words whose meanings can be either skewed or understood differently—or not at all—in another language. At the very least, it emphasizes the danger of taking for granted that English words have universal meanings and uses.

Here we can look again to Holberg's Jean de France: "If only she did not speak Danish." Two of the most dangerous attitudes when working in local markets are disrespect for local languages and lack of attention to them. These attitudes create distance from content users who might even be hesitant to reveal their ignorance or might build up outright resistance. While technical communicators and other content developers might hide their lack of understanding and pass the buck to the customers, the resulting lack of understanding and misunderstanding can easily spill over to products so that they are seen as "flawed," "substandard," or "defective." The call for technical communicators working in both global and local contexts is to heighten their awareness of the effects of a lack of translation and localization of text as well as a possible lack of adaptation of products.

Respect for the audiences of technical communication material calls for more attention to the routines of global communication that needs to travel to different locales. Crossing the barrier is not just a matter of asking a local salesperson to go through the material. Very often salespeople are too busy to take the time to localize something that takes an extra effort—or worse, they will not admit that they do not understand the material themselves and just pass the buck to the next link in the distribution chain. What needs to be done in sales and distribution channels is what is routinely done in the case of instructions: a usability test, where usability means "how something is understood." Only then can a marketer—globally or locally—decide whether something should be localized for the sake of comprehension or should be left in English as a foreignizing element for status, educational, novelty, or other reasons. The wrong choice among the different options might mean a customer walking away or maybe not showing up in the first place.

Conclusion

To return to the first research question that guided the study reported here—do non-native (L2) English users define and use English professional loan words in the same way as native (L1) English users—the answer is both yes and no. It depends entirely on the word and the context in relation to the following factors.

Loan Words and Associations

One of the biggest problems in relation to using loan words is the transfer of complete understanding of the same word in the same manner. The underlying idea and root of the word might not exist in the L2. We know this from many classic examples: Danes relate the word *computer* only to the machine, not to the function of the machine *to compute*. (This is because the verb "compute" does not exist in the Danish mental lexicon.) Marketers introducing a loan word for a non-English market should therefore be aware that the underlying idea of loan words does not necessarily accompany the word into the other language.

The Fallacy of Phonological or Morphological Resemblance

Another reason for understanding a word differently is the phonological or morphological resemblance with another word. We saw this in the examples *phishing/fishing* and *sneakers/snickers* and possibly in the morphological resemblance *phishing/piercing.*

The Fallacy of Domain-Specific Language

A third fallacy is the use of domain-specific language in communication with users. The most negative consequence for a content creator might be that people either abstain from inquiring about a product because they feel ashamed of their ignorance or feel annoyed that they are left out of communication. That problem may be ascribed to pseudo-context: Saying "a bicycle with *full suspension*" does not give any added information if the buyer does not know *full suspension*.

This factor leads us to the second research question: In which ways do non-native (L2) English users use professional loan words differently from native (L1) English users?

The Primary Understanding of a Word

The primary understanding of *full suspension* in a Danish mental lexicon connects with being temporarily laid off from a job. The technical meaning of being equipped with springs does not exist in the Danish lexicon, where the word *affjedring* [spring effect] is used to express *suspension*. Whether lack of knowledge, laziness, or prestige is accountable for the non-translation is an open question here. A trickier example became evident from the word *streetsmart*, where the primary meaning in English is to be streetwise, but the primary Danish meaning is wearing modern, hip, young clothing, maybe as a result of the different weighting of the primary meaning of the word *smart* itself, which is *good-looking* in Danish but *clever* in American-English.

The Contextual Link

Another point is the importance of the contextual link to the society in question. A couple of the words in this study's database suggest exactly that. The most obvious example is the word *sneakers*. The slogan *snup en snickers* [pick a Snickers] was rooted in an advertising context, thereby triggering images of a chocolate bar for the word *sneakers* even in a textual context where the meaning was shoes.

Generation-Specific Words

My respondent group was young people aged 17–24. For some of the words, different responses might appear if another age group was asked. I picked this age group because the common perception is that young people understand English better than their parents' generation—and thus also English loan words. This point is still open for research, but the general understanding and use of English loan words among young people leave room for improvement. Also, the rate of understanding and use by a comparable American age group for the same words would be interesting to see.

Resistance to Loan Words or Status of Using Loan Words

The issue of resistance to, or status gained from, using loan words did not emanate clearly from the responses, probably because the questionnaires did not include prompts for that. However, the voices of resistance, though expressed by few, were quite strong; moreover, they depended on the word in question.

Streetsmart was again one of the words to which resistance was voiced in the form of typical comments, such as "the word does not function in the Danish language" and "an empty marketing word," whereas some voices expressed status or young street life, such as "cool," "street respect," and "to be street," with this last phrase being a pseudo-loan.

Replicating the Methods of Research

Technical communicators and other content creators who would like to replicate my methods might find the process somewhat cumbersome. To obtain quick results, one might confine methods to the combination of *intuition* with the *real context*. With this approach, marketers could still check their overall use of loan words in a given context while coupling them with an initial understanding based on the respondent's existing mental lexicon.

In this study, I may have overlooked many reasons for using loan words. In a technical communication context, however, what is most important? To be a first mover or to get hold of the mass market—thus challenging customers with the leading edge of the language change and use or playing it safe? Or the choice may be even simpler: How much added revenue (and added market share) does a company get for each dollar spent on translation and localization, compared to passing on loan words?

Notes

1 With statistical advice and evaluation from Carl-Johan Nielsen
2 *par Di*; should have been *par Dieu*, in the same way as by "God." The phrase was normally used as a sentence enhancer or a small break in the sentence. *Vimmelskaft* is a street name in the old quarter of Copenhagen. It is a part of the main pedestrian street today and has gotten its name because of the crooked nature of the street, which is probably the intended metaphor.
3 Translated from Dano-Norwegian playwright Ludvig Holberg's comedy titled *Jean de France eller Hans Frandsen*. Excerpt in original Danish: Pierre: "Mad. Isabelle har dog et kiønt Ansigt." Jean: "Ja, det er kiønt nok, men det er dog saa Dansk, c'est une visage a la Danois, a la Vimmelskaft par Di, dog jeg kunde nok venne mig til Ansigtet, dersom hun icke talede Dansk."
4 Free word association has been used within both psychology and linguistics to compare differences between the primary language (L1) and the second language (L2) lexicon (Zareva 2007), and as a method to analyze word class influence (Nissen and Henriksen 2006). To label my findings according to any existing system would probably be wrong, because most existing research is based on frequent words in the mental lexicon (Higginbotham 2010), not loan words.
5 Back-correction could be a problem if the contextual questions were given on the same sheet as the word association questions, because people tend to understand words better in context. For this reason, the word association questions were kept apart from the contextual questions.
6 I use the term "semi-false friends" for words that exist in two languages but either with slightly different meanings or with different first-association uses of the word.

References

Collins English Dictionary. n.d. "streetsmart." Accessed December 30, 2014, and July 27, 2018. www.collinsdictionary.com/dictionary/english.

Cremer, Marjolein, Daphne Dingshoff, Meike de Beer, and Rob Schonen. 2011. "Do Word Associations Assess Word Knowledge? A Comparison of L1 and L2, Child and Adult Word Associations." *The International Journal of Bilingualism* 15, no. 2: 187–204.

Dalsgaard, Marianne (2015). "Modeshow på museum er *sooooo last season*." [Fashion show at museum is sooooo last season.] *Historie*, Nr. 1. Copenhagen: Bonnier Publications A/S: 6.

Dansk Sprognaevn. n.d. Retskrivningsordbogen. www.dsn.dk

Davidsen-Nielsen, Niels, Erik Hansen, and Pia Jarvad, eds. 1999. *Engelsk eller ikke engelsk. That is the question. Engelsk indflydelse på dansk*. [English or not English. That is the question. English influence on Danish.] *Dansk Sprognævns Skrifter*, Nr. 28.

Den Danske Ordbog. n.d. Accessed August 23, 2013 and July 29, 2018. http://ordnet.dk/

Ejsing, Jens. 2009. "Flertal sikrer forskning på dansk." *Berlingske Tidende* (Copenhagen), January 19.

Eriksen, Thomas Hylland. 1992. "Linguistic Hegemony and Minority Resistance," *Journal of Peace Research* 29, no. 3: 313–332.

Fernández, Rosa M. 1990. "Actitudes hacia los cambios de códigos en Nuevo Mexico: Reacciones de an sujeto a ejemplos de su habla." In *Spanish in the United States: Sociolinguistic Issues*, edited by J. J. Bergen, 49–58. Washington, DC: Georgetown University Press.

Filipovic, Rudolf. 1996. "English as a Word Donor to Other Languages of Europe." In *The English Language in Europe*, edited by Reinhard R. K. Hartmann, 37–46. Oxford: Intellect.

Haberland, Hartmut. 2005. "Domains and Domain Loss." In *The Consequences of Mobility*, edited by Bent Preisler, Anne Fabricius, Hartmut Haberland, Susanne Kjærbeck, and Karen Risager, 227–237. Roskilde, Denmark: Roskilde University.

Heidemann Andersen, Margrethe. 2002. "Engelsk i dansk: Sprogholdninger i Danmark. Helt vildt sjovt eller wannabeagtigt og ejendomsmæglerkækt?" Ph.D. dissertation, University of Copenhagen.

Higginbotham, George, 2010. "Individual Learner Profiles from Word Association Tests: The Effect of Word Frequency." *System* 38, no. 3: 379–390.

Holberg, Ludvig. 1723. *Jean de France*. http://wayback-01.kb.dk/wayback/20101103142713/http://www2.kb.dk/elib/lit//dan/holberg/komedier/jean.dkl/a3.htm

Holberg, Ludvig. n.d. "Jean de France eller Hans Frandsen." Ludvig Holberg Skrifter. Accessed December, 10, 2014. http://holbergsskrifter.dk/

Jarvad, Pia. 1999. "Nye ord—Ordbog over nye ord i dansk 1955–1998." *Gyldendal*, 1. udgave.

Laurén, Christer, Johan Myking, and Heribert Picht. 2002. "Language and Domains: A Proposal for a Domain Dynamics Taxonomy." *LSP and Professional Communication* 2, no. 2: 23–30.

Lønsmann, Dorte. 2009. "From Sub-Culture to Mainstream: The Spread of English in Denmark." *Journal of Pragmatics* 41, no. 6: 1139–1151.

Matus-Mendoza, Mariadelaluz. 2002. "The English Lexical Loan: A Class Marker." *Journal of Hispanic Higher Education* 1, no. 4: 329–337.

Merriam-Webster Dictionary. n.d. "loanword." Accessed December 8, 2014, and July 27, 2018. www.merriam-webster.com.

Nissen, Henriette Bagger, and Henriksen, Birgit. 2006. "Word Class Influence on Word Association Test Results." *International Journal of Applied Linguistics* 16, no. 3: 389–408.

Oxford English Dictionary. n.d. "boardshorts." Accessed December, 30, 2014, and July 27, 2018. www.oed.com.

Oxford Learner's Dictionary. n.d. "loanword." Accessed December, 30, 2014, and July 27, 2018. www.oxfordlearnersdictionaries.com/definition/english/loanword.

Preisler, Bent, 2003. "English in Danish and the Danes' English." *International Journal of Society and Language* 159: 109–126.

Rasmussen, Iben Lehmann. 2008. "Domænetab: engelske låneord i Würth Danmark A/S produktbog. Hvad betyder de egentlig?" Graduate thesis, Handelshøjskolen i Århus [Aarhus School of Business].

Rathje, Marianne. 2012. "Luder og laban. De grimmeste ord i to generationer." *Nyt fra Sprognævnet*, 2. juni.

Selsøe-Sørensen, Henrik. 2004. "Ord for det, men ikke på dansk." *Politiken*, 5. august.

Svane, Christina, 2010. *Engelsk i dansk og risikoen for domænetab*. Master's thesis, Copenhagen Business School.

Zareva, Alla. 2007. "Structure of the Second Language Mental Lexicon: How Does It Compare to Native Speakers' Lexical Organization?" *Second Language Research* 23: 123–153.

7

TECHNICAL COMMUNICATORS ON TRACK OR LED ASTRAY!

Using Quality Standards in Practice

Birthe Mousten and Dan Riordan

Introduction

> A company drafted a Romanian/German bilingual contract, stipulating the delivery of 3,000 cars, half black, half yellow. Obviously, the firm wanted 1,500 yellow cars and 1,500 black. What happened (as a result of mistranslation) was that all products had yellow hoods (half yellow) while the rest was black (the other approximate half).
>
> (*Cismas 2010, 492*)

Technical communication instructors increasingly need to familiarize students with the demands of working in a globalized environment that increasingly involves individuals translating, revising, and reviewing texts within cross-cultural virtual teams (CCVTs)—groups of individuals who use online communication technologies to interact from different locations around the world (Herrington 2008; Huc-Hepher and Barros 2016; Humbley et al 2005; Maylath, Vandepitte, and Mousten 2008; Mousten 2008).[1] While many instructors routinely prepare their students to write for translation and internationalization, few focus on the translation standards in other parts of the world (Metzger 2011). Yet variations across such standards can have important implications for documentation requirements in international contexts. The contents of these standards and how they differ can therefore be essential to working effectively in a global context.

In Europe, the European Union's (EU) Machinery Directive (Directive 2006/42/EC replaced by Directive 2006/42/EG in 2017) specifies that, "All machinery must be accompanied by instructions in the official Community language or languages of the Member State in which it is placed on the market and/or put into service" (European Parliament and Council of the European Union 2017, point 1.7.4). Such requirements call for a high level of translation quality, and to address quality in trade and industry (T&I), most European countries have agreed on a translation standard that delineates a process, which all translation service providers (TSPs) should follow when translating documentation. The objective of translation and documentation standards is to avoid misunderstandings, such as the half-and-half car order noted in the beginning of this section and to provide a process for more easily identifying and addressing problems during documentation drafting over the accompanying translation to the actual market.

The translation standard, known as EN 15038, was drafted in the early part of the millennium, adopted for use in 2006 (CEN 2006a), and updated with EN 17100 in 2015 (CEN 2015). A range of businesses now uses this standard as a guide to cover the complete work process from translation of source text (i.e., original text to be translated) to the target market text (i.e., final translated and adapted text). The 2015 standard further includes project management as a specific skill for people and also includes work phases, such as IT-aided phases, as well as post-editing (editing of machine-translated texts) as a human process. Through their marketing activities, companies worldwide are generally affected by the standards and may find it useful as a quality assessment and enhancement tool to define and assign responsibility.

The standards (EN 15038 and EN 17100) mirror the translation process commonly used in T&I, and it links three work processes: translation, revising, and reviewing, as well as an overall project management. As the three-party work process implies, preparing documents for the target market or target country involves collaboration between team members in different markets and with different languages: the so-called cross-cultural, virtual teams (CCVTs). They comprise a translator and a reviser from the source country and a reviewer from the target market country. Many technical communication students have probably had little exposure to the process and the demands of working in teams guided by such a process.

In technical communication translation practices, crowdsourcing has exploded, and machine translation (MT) and post-editing are already common practices. Editing and post-editing, as well as automation of the translation process, have increased the requirements for revision and editing practices as well as automatic text-processing practices and post-editing. In turn, the need to understand and handle the basic interaction between the translation-revision-review phases will be requisite for any technical communicator.

Since 2003, this three-party-process has been incorporated in technical writing and translation programs among students and instructors at Aarhus University (Denmark), the University of Wisconsin-Stout (USA), and later at North Dakota State University (USA), as well as many other universities across four continents. In such cases, groups of students from different nations collaborated on technical writing and translation projects. These projects were designed to reflect the scenario of marketing products in the United States or the European Union, and the objective was to help students understand how such processes could affect professional communication practices. This chapter examines the lessons learned from these projects. With this focus in mind, the authors of this chapter address the overarching question: In trading with European nations, what do technical communicators need to know about the processes delineated in the European translation standard?

Overview of the Theories and Processes

Today, approaches to translation studies could be based upon a range of theories, depending on the translation's use. Within this context, two important theories for understanding translation processes are relevance theory and translational action theory.

Relevance Theory

Relevance theory has moved the focus of the translation process away from the source text (i.e., the text to be translated) and towards the target market text (i.e., the final, translated text). First presented by Sperber and Wilson (1995), Gutt (2000) connected relevance theory to translation. By focusing on the sender-to-receiver dynamic, Gutt switches *translation* from meaning "two language versions of the same text" to meaning "two texts with similar purpose and understanding." In this case, the source-text version (original) is designed for individuals in one language and target group, namely the original source-text audience. By comparison, the target-market text version (translated text) is designed for the members of another language and target group, namely the target-text audience.

The old view of different text versions as identical texts in different languages has changed to focus on the participants' ability to see the source text as a hypothesis, which has to be tested, proven, changed, or even rejected through the work process (Chesterman 2016). According to this shift, the tasks of the translator increasingly become a matter of analyzing the audience's (the target market's) comprehension of the topic and the text. In relevance theory, this analysis process works through assumptions, which are based on things or ideas that the audience comprehends without explanation. The translator must learn to identify and assess the relevance of assumptions (i.e., prior knowledge, namely what the audience already knows about the topic). Two rules guide this analysis process:

- An assumption is relevant in a context to the extent that its contextual effects in this context are large.
- An assumption is relevant in a context to the extent that the effort required to process it in this context is small (Sperber and Wilson 1995, 125).

Accordingly, as they create the new text during the process of translation, CCVT-members must analyze any change in relation to those two rules concerning assumptions. According to this approach, some textual statements have to be expanded so the new audience will realize the importance of those statements. This process is also known as explicitation (Becher 2011).

Consider the following example. To convey the *Washington Post* news headline, “The GOP is buying the House. Literally” (Milbank 2018) to others, any translator or political commentator would have to first make explicit that “GOP” is short for “Grand Old Party,” or rewrite the term to say “Republicans.” Second, to be on the safe side, the commentator or translator would have to expand “the House” to “the House of Representatives,” to make sure that readers might not mistakenly think the term meant the White House. In these cases, texts have to be expanded, for the commentator or translator cannot assume that readers will correctly understand the meaning of “GOP” or “the House” in this text.[2]

Some statements, however, can be deleted or shortened without the loss of meaning. This process is known as implicitation (Becher 2011). To see how explicitation and implicitation work, consider the following example from a British newspaper: “David Cameron has attempted to explain why he decided to lean in for a ‘selfie’ with Barack Obama and the Danish prime minister Helle Thorning-Schmidt during Nelson Mandela’s memorial service in Johannesburg yesterday” (Selby 2013). Depending on where in the world this information is shared, there is a difference as to which of the persons in this example should be mentioned in terms of political office. Or, to say it another way, how the text will be explicitated or implicitated. If the excerpt were published in the United Kingdom, the Danish prime minister would need to be identified by office, as UK readers might not know her position. Obama and the British PM would not need to be identified by office, based on the assumption that UK readers know who these individuals are and what offices they hold. In other parts of the world, the need for explicitation or implicitation might change, depending on how well readers know who the individuals mentioned are.

According to relevance theory, the audience’s efforts to understand the text must be kept low at all times (i.e., readers should not have to expect too much mental effort to determine what is conveyed in a message). For this to be achieved, texts need to be constructed to ensure that the audience’s contextual relevance is large. In other words, writers and translators need to take care to write by assessing what readers can be assumed to know about the topic. In the previous example, many newspapers included an explanation of the word *selfie*, which was a relatively new term at the time when the news was reported, but at the time of this book’s publication would be obvious knowledge.

Translational Action Theory

The second theory, known as translational action, defines communication as that which fulfills the needs of others (Holz-Mänttäri 1984; Risku 1998; Breedveld 2002; Reiss and Vermeer 2014; House and Loenhoff 2016). Translational action is therefore an instrumental part of what can be termed process action or the freedom to interpret and change elements to convey meaning. In most business contexts, process action involves a variety of professions to fulfill the communicative purpose and thus meet the communication needs of others. Katan (2009) has investigated professional translators' work processes and found that their loyalty extends to the source text, the target text, and the client. Along the same vein, translators have voiced responsibility to the client, references to colleagues, and quality assurance as part of their general obligations (Ehrensberger-Dow 2014).

According to these perspectives, no one can request that a translator—given any odd text in hand—should have a full and total overview of the contextual differences between two texts in two languages and the processing efforts of the two audiences placed in different cultures. The combination of relevance factors, translational action, explicitation, and implicitation—to mention just some of the factors involved—places huge demands on technical communicators and translators. For this reason, it is important to understand how such factors affect professional communication and translation practices today.

In professional contexts, a major shift has taken place since the 1980s. Today, translators rarely work as single-language vendors (SLVs): that is, one sole individual doing all of the related translation work on a project. Rather, translators now tend to work on a text in networks or as part of a translation or localization team. Moreover, such teams have to function not just in one new locale but in several new locales at the same time (Mousten 2008, 41).

Additionally, these teams should be bilingual, as noted by Brunette, Gagnon, and Hine (2005), who have reviewed monolingual revision compared to bilingual revision. They found that although the translations turned out to be better in bilingual revision, both over-revision and under-revision existed. (Over-revision is introducing errors in the translation and under-revision is making too few revisions.) It turned out that monolingual revisers sometimes introduced more errors into a translated text than the number of errors that they corrected (Brunette et al. 2005, 35–36). The Brunette et al. study thus indicates the need for participants in translation teams to understand the rules of assumption and the methods for explicitating and implicitating.

Revision and Review as Work Areas

Self-revision is the translator's revision of his or her own text and the duty to do so (Robin 2016). It has been researched through think-aloud protocol studies and keystroke logging (Breedveld 2002, Jääskeläinen 1999, Jakobsen 2006) as well as eye-tracking (Schrijver 2014). These studies have provided insights into the translator's translation and revision processes; however, they examine only one person working alone. As a result, they cannot sufficiently shed light on revision processes made by another person. Because of this disconnect, the EN 15038 or EN 17100 standards specify that a translation team have clear service specifications for the translation project. Doing so allows the team's processes to be streamlined.

When a person other than the translator has to revise a text, some preconditions must be established to avoid confusion based upon which standards to use to guide the process. Within this context, Hayes (2004) has suggested that team members perform three tasks:

- Task 1: Establish the goals for revision: When and how will the text be used?
- Task 2: Determine the scope of the revision: What is expected of the reviser in terms of work, for instance, correction of the language only or also the text's system-bound references (i.e., references that would be clear to native readers of the source text but not to readers of the target market text)?
- Task 3: Agree on the procedures to be used: Which systems of correction will individuals use?

Defining these three tasks helps teams of one or more translators, revisers, and reviewers distribute and cooperate on the whole process. This defining process would include everything from drafting the translation to publishing a text in another language and culture (Hayes 2004). These preconditions must be included in the service specifications created for the current project (EN 15038 and EN 17100).

Specifications in the EN 15038 and EN 17100 Standards

With these ideas in mind, let us turn to the European standard EN 15038 drawn up by the Technical Committee CEN/BT/TF 138. The standard was approved by the European Committee for Standardisation (CEN[3]) in 2006 and was subsequently approved by its CEN members (most European countries). The standard exists in three officially approved versions: English, French, and German (EN 15038), with subsequent translations into other languages, including Danish (DS/EN 15038). Officially, EN 15038 was withdrawn and replaced in 2015 by EN 17100, but the same contents are included with added specifications for project management. As the examples here go back to before the time of EN 17100, and the project worked on the basis of the EN 15038 standard, this standard has been included as the basis for the analysis of the examples in this chapter.

The EN 15038 standard is directed toward translation service providers (TSPs) and establishes specifications (see Figure 7.1) for the members of a cross-cultural virtual team (CCVT), which is comprised of the translators, the revisers, and the reviewers. The processes and powers of each person or professional group in a CCVT are also shown in Figure 7.1.

The standard's specifications for each Content Provider (CP) are visible in the thought bubble at the top of each person in the figure. Moreover, the language competencies and job responsibilities are shown in each face, and the boxes at the bottom illustrate how the texts can draw on and adopt cultural elements from the sender and receiver cultures respectively. The overlapping faces and thought bubbles symbolize that cooperation between the people in the process are necessary for a high-quality, negotiated result.

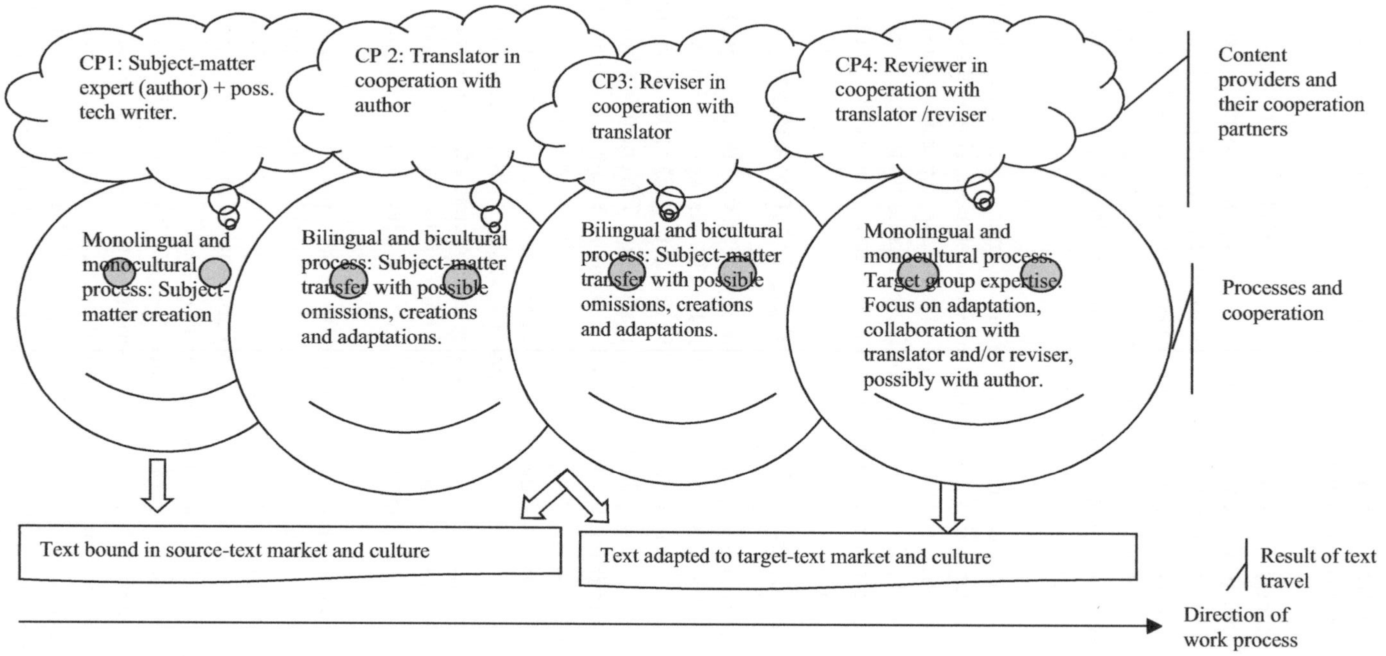

FIGURE 7.1 Content providers, work process, and results

Responsibilities and Competencies of the Translator

Standard EN 15038 specifies the work responsibilities of the translator as follows:

> The translator shall transfer the meaning in the source language into the target language in order to produce a text that is in accordance with the rules of the linguistic system of the target language and that meets the instructions received in the project assignment.
>
> *(CEN 2006, 11)*

Noteworthy here is that the project assignment defines the freedom of the translator to rely on either the source-text content or the target-market requirements in the translation process as defined by the service specifications. Some project assignments will require the translator to provide an old-fashioned translation in which the source text is rendered more or less word for word. Others will require a translation that reflects the concepts of relevance theory or translational action theory.

Responsibilities and Competencies of the Reviser

The role and work responsibilities of the reviser are specified in the EN 15038 standard as follows:

> The reviser shall be a person other than the translator and have the appropriate competence in the source and target languages. The reviser shall examine the translation for its suitability for purpose. This shall include, as required by the project, comparison of the source and target texts for terminology consistency, register and style.
>
> *(CEN 2006, 11)*

The work of the reviser therefore consists of more than checking the initial translation for correctness. It also involves the very important job of looking at the suitability for purpose in the translated text. Thus, the work of the reviser has a Janus face. One side of the face is a source-text and target-market comparison; the other side of the face examines the translation for its suitability of purpose in the target market. Those two requirements are often incommensurable, because many suitable translations call for omissions, additions, and replacements in a text: in other words, effective use of relevance or translational action theories.

Responsibilities and Competencies of the Reviewer

The work responsibilities of the reviewer are specified in the EN 15038 standard as follows:

> If the service specifications include a review, the TSP shall ensure that the translation is reviewed. The reviewer shall carry out a monolingual

> review to assess the suitability of the translation for the agreed purpose and recommend corrective measures.
>
> *(CEN 2006a, 11)*

Notice that the role of the reviewer is initiated by the word "if." This means that this step in the process is not obligatory. Rather, it must follow what is called for in the service specifications. This contingency means the step is subject to contractual obligations and not to the specifications laid down in the standard.

Notice too that the reviewer works monolingually with the same responsibility as that of the reviser, namely to achieve suitability for the agreed purpose. Moreover, the role and responsibility of the reviewer are also to recommend corrective measures—or to suggest changes to the context which will fulfill cultural purposes or marketing decisions. If the reviewer suggests changes, the text will have to go back to the translation/revision stages for adaptation. This is because the requested competence of the reviewers is oriented only toward the target audience and thus is unconnected to the origins of the text. This process means that unintended or cross-purpose changes can be made for the target audience. An example of this approach is provided later in the chapter.[4]

Negotiation or Feedback

A vague point in the EN 15038 standard is the process for resolving possible changes in the text. For instance, should they be addressed through dialogue and negotiation in the process chain, or as collected feedback to be followed after the process is complete? The EN 15038 standard specifies that the translator shall make any necessary amendments, and the TSP shall

- take the reviser's and the reviewer's recommendations into account and
- ensure that any necessary corrective measures are implemented. (CEN 2006a, 11–12)

This specification thus makes several processes possible. For instance, a TSP could review the recommendations, then take corrective action, or ignore those recommendations. Again, the processes of negotiation have not been specified in the standard but have to be laid down by the translators' and technical communicators' quality assurance systems.

There is no doubt that the TSP has primary responsibility for the project. When it comes to implementing corrections, the terms "take into account" and "necessary" mean that the TSP has some leeway in discussing with the translator/reviser team the relevance of the changes suggested. Because the processes of implementing corrections may be problematic in the feedback and negotiation phases, negotiation processes ought to be considered in the detailed service specifications.

An Experiment with Teamwork and Work Flow

CCVTs are now everyday realities of the workplace (Anawati and Craig 2006). To replicate this reality in the classroom, the pilot projects reported on here were set up as CCVTs where the translation and revision work took place in one country, Denmark. The reviewing phase, in turn, took place in the United States. The team members came from Danish and US classes, and the composition and configuration of participants in the project reflected the organizational patterns in industry.

Organizing International Project Teams

For this project, the organizational teams were divided between two classes in two countries: a master's-level translation class in Denmark that focused on technical translation to and from English and a bachelor's-level senior capstone seminar for technical communication majors in the United States. To maximize the learning context, the two instructors (the authors of this chapter) formed the CCVTs typically with two Danish students and two US students. The bilingual Danish members translated and revised a text, which their US partners then reviewed, having been instructed to assess the text for the purposes of possibly publishing the resulting translation in the northern United States. The underlying idea was that a Danish company wanted to introduce and possibly market a product in the United States and thus needed to publish and distribute some material about its inventions, work, and products for members of a given customer group. The process of organizing the workflow to realize this scenario created new challenges for the language work across borders and across the Atlantic Ocean. This latter factor meant that they had to account continually for a seven-hour time zone difference. Because of the time difference, communication among the team members occurred almost exclusively via email, while the texts that teams translated, reviewed, and edited were typically displayed in Microsoft Word.

In partnering to examine these translation-related processes, the two instructors delegated as much responsibility and work to the CCVTs as possible. Using this approach thereby mirrored the process delineated in the EN 15038 standard. Additionally, the instructors took on roles comparable to project managers (whose inclusion would be an EN 17100 recommendation) in relation to the teams of students, who, with their respective bilingual and monolingual backgrounds, took on the tasks of translators, revisers, and reviewers.

Service Specifications in the Project

When creating the work teams, the instructors provided students with service specifications that included three obligatory processes, as shown in Figure 7.2. To start, a Danish team member chose as a source text an article from a technical

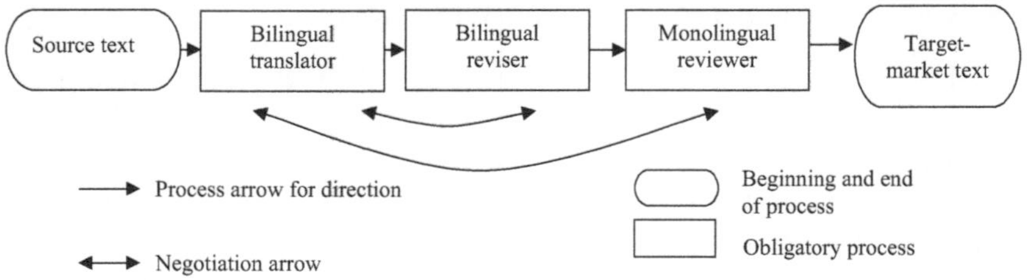

FIGURE 7.2 Work process including negotiation process

magazine and translated it. Then, a Danish classmate revised and proofread it. The team members in Denmark then sent the revised text (the target-market text) to a team member in the United States for further review. During this review phase, the US students suggested and negotiated changes with the Danish translators. The text resulting from these interactions was considered the final, translated target text. Schematically, the process can be shown as follows:

Figure 7.2 illustrates text travel: the processes and decisions involved in transforming a source text in a source language into target-market text in the target language. In producing the final translation, the members of the student teams adhered to the obligatory double-bilingual process (translation and revision) in the EN standard. Additionally, the voluntary monolingual review process in the EN 15038 standard, as carried out by the US students, was included as an obligatory part of the work process. For this project, the review process was entirely monolingual, as it all took place in English. As Figure 7.2 indicates with its negotiation arrows, the process moves both forward and backward until the target-market text is declared final.

Texts and Translations

Because the class project involved translating texts from Danish trade magazines, specific Danish cultural factors and conditions surfaced as the students translated articles into English and adapted them for appropriate trade magazines in the United States. Many of the magazines covered a wide range of energy issues, both conventional and renewable. The translated texts provide US audiences with information about developments and projects in other places in the world. Hence, the service specifications for the project entailed that the translated text should simultaneously be source-culture-oriented but also target-market-adapted, so that the audience would comprehend fully.

In this example, source-culture-oriented means that the text clearly originates in a culture other than the one in which it is read, and target-market-adapted means that the foreign elements in the text are explicitated or adapted to the target. The setup of the student CCVTs thus closely resembled the CCVT constellations in many workplaces: It operated in an environment of cooperation across professions, borders, and time zones.

As in many workplace CCVTs, the majority of the actual work took place in the two countries, even though the process was collective. This process is actually an innovation. In previous studies on this type of translation work, the direction of work started with domain specialists/subject-matter experts and proceeded to the technical writers and then the translators. In contrast, in the project described here, the translators started the text production by translating into English an already published text in Danish, and then the US reviewers suggested corrections. The result was that the native-English-speaking reviewers were much more consultative than has been commonly seen in earlier projects when the source text was English. (The effects of this reversed direction of text travel have been described in Humbley et al. 2005 and Maylath et al. 2008.)

Empirical Examples from the Work Process

The following examples from the CCVT collaborations illustrate some of the changes made during the work process, by whom, and to what degree. In these examples, the abbreviation "T" means *Translated Text*, "RV" means *Revised Text*, and "RW" means *RevieWed Text* (see Examples 1 and 2). The underlined passages (see Example 3) show if and where changes were made in the translation/revision stage, and bold text (see all examples) signals changes made during the review stage.

Example 1: Questionable added value of review

T + RV: As one of the biggest players in the field, the Danish company Chr. Hansen A/S has automated entirely its production of enzymes and bacteria cultures to ensure the quality.

RW: As one of the **most respected players** in the field, the Danish company . . .

In Example 1, the text introduces the company, a wholesaler of additives to industrial customers. This example reveals a change in content, altering the semantics: "the biggest player" has been changed to "the most respected player." In many cases, this change could well be an improvement to the text, as it could be assumed in the contextual situation (i.e., for the target market) that it is better to be "respected" than to be just "big." However, in this particular contextual situation (i.e. in the source's context), the change is questionable. Why? Because a company dealing in enzymes, additives, and bacteria cultures does not necessarily want to be well known to the public. Both the customers and the products are industrial. More importantly, mention of additives, enzymes, and bacteria cultures can inadvertently raise alarms among members of the public even if their uses promote the general good. However, if neither the company nor its customers wish to be in the public eye, then there is no basis for being well-respected. Thus, in this example, the reviewer's semantic change with an eye to the target market fails to take into account the company's position in the source culture as well as in society.

In Example 2, the text is excerpted from instructions on how to recharge a battery. In this example, an element is omitted to get rid of information that the reviewer considers superfluous in the target-market language and culture.

Example 2: Review fails

T + RV: By storing, a slow chemical reaction will occur reducing the amount of energy stored in the battery. The relation between energy and weight (or volume) is called storage density.

RW: By storing, a slow chemical reaction will occur reducing the amount of energy stored in the battery. [—]

The reason that the reviewers gave for omitting the second sentence was that the article was intended for a lay audience. For this reason, the reviewers argued, complex concepts that require a specialist's technical knowledge can be omitted. In many texts, this argument would make sense. However, in this instance, the concept of storage density was used later in the text, and the definition was required for readers to understand the details and examples provided in the text. In other words, the omission might be appropriate when one considered a lay audience, but the contents revealed that the source text was, in fact, aimed at a more specialized-knowledge audience. The concept of the storage density was included so that readers could understand other elements in the text further on. For this reason, the definition of storage density needed to be included to allow for full comprehension of the text to its final pages.

In Example 3, the complete process—from translation to revision to reviewing—shows improvement in many respects. However, it also has one erroneous correction. Explicitation and changes in sentence structure are applied.

Example 3: Successful translation, revision, review process—with a minor exception

(In this iteration, the underlined phrases are revisions, in the form of explicitations, to the original translation.)

T + RV: While the <u>Danish center-right</u> government tries hard to renew its election promises, and the <u>internal struggle within the major opposition party,</u> the Social Democratic Party, seems to have been revived, Ulrich Pade of the Center Democratic Party is trying to introduce into the debate the subject of building a wave power production plant on Lolland-Falster, <u>an island region with high unemployment rates.</u>

This example of translation and revision in the process shows that the translator and the reviser (who wrote the underlined phrases) together added many explicitations to make the text target-market-oriented. Adding context (the underlined phrases) results in demanding less processing effort on the part of target-language readers. To achieve this result, the reviser supplemented the text with explicitations that Danish readers of the source text would add on their own subconsciously. For instance, the source text's *regeringen* ("the government") is explicitated into "The Danish center-right government" and *krig* ("war") becomes "struggle" in the reviser's effort to change the tone of the language to something more palatable to the target audience.

the government	}→	The Danish center-right government
war between members of the Social Democratic Party	}→	Internal struggle within the major opposition party, the Social Democratic Party
Lolland-Falster	}→	Lolland-Falster, an island region with high unemployment rates

In the subsequent review stage, still more changes were made, appearing here in bold:

> RW: **While** the Danish Center-right government tries hard to renew its election promises, ~~and~~ **the internal struggle** within the major opposition party, the Social Democrats ~~ic party~~, seems to have been revived, Ulrich Pade of the Center Democratic Party is trying to introduce ~~into the debate~~ the subject of building a wave power production plant on Lolland-Falster, an island region with high unemployment rates.

The purpose of these changes was to make the language flow more naturally and to make the text suitable for the target audience. For example, changing "to introduce into the debate the subject of" to "introduce the subject of" undoubtedly heightens the quality of the text and increases the fluency of the text.

The second, and most impressive, review change was to switch "the Social Democratic Party" to "the Social Democrats." This minor change reflected the fact that the Social Democratic Party some years ago changed its name to the Social Democrats—a change so small that it was not caught first by the two bilingual translators from the source culture. Apparently, the monolingual reviewers must have accessed the website of the political party for background knowledge.[5]

The third change, which seems to be just a minor syntactic change, turned out to be a major semantic change and was not acceptable. In this case, the change was from "while the Danish . . . , and the internal struggle . . . , Ulrich Pade" to "While the Danish . . . , [___] the internal struggle . . . , Ulric Pade." The Danish semantics points to three issues: While the government struggled with one issue and the opposition party struggled with another issue, Ulrich Pade introduced a new, troublesome issue. In the Danish text, "while" introduces two contrasts: the government plus the opposition on the one hand versus Ulrich Pade on the other. By omitting the "and," the reviewed text makes the contrast between the two political parties, not between existing technology and new technology. The topic of the text thus shifted. The observant reader also notices that the syntax falls apart.

Evaluation of Process Results

Overall, the results of the translation-revision-review process were as diverse as the texts and the students involved. There were good translation-revisions and bad reviews just as there were erroneous translation-revisions and good reviews. Fortunately, in most of the places, the process added value all the way through the process.

In the example (1) of the additive-producing company, the review failed, because the relevance of the semantic change in the contextual situation was doubtful. As the company was not well-known (one of the big Danish companies not recognized by the general public) and did not want to be so, it would make the target audience assume that something is true when in fact it is not. Had the change stood, target-market readers unfamiliar with the context in Denmark might easily conclude that the company is well-known. Ignorant of the Danish context, they would not be aware that the assertion is wrong. Relevance theory would thus require rejecting the change because of the contrast with reality and the misleading of the audience.

In the battery example (2), with its definition of a difficult concept, the review failed because the intra-contextual relevance was high; in other words, the reader needed the definition to understand explanations made later in the text. Therefore, the information—specifically the need for the term *storage density* for batteries—was relevant and needed. Also, the efforts on the part of the reader to fully comprehend and process the text would require high efforts, with little likelihood of success unless the reader undertook his or her own research outside the text. Relevance theory would again require rejecting the change, because the relevance of the omitted information was high and needed for complete comprehension of the text.

In the wave power example (3), the translation, revision, and review processes were somewhat successful because they added information—for instance, the political implications of an island with high unemployment and thus the need for development. The explication therefore eased the readers' understanding. The contextual effects were thus large, and the effort to process the information small. A minor hitch was the shift of focus away from the central topic to other problematic issues.

Evaluation of Revision and Review Parameters

In addition to the processes already described, the EN 15038 standard specifies some quality assurance and work process methods. These items allow both effective evaluation of the work process and the basis for training team members in good practices for the translation-revision-review process. One scholar who has identified criteria for evaluating work processes is Brian Mossop.

In the students' project, the service specifications did not include any detailed principles for translation, revision, or reviewing of the texts. This was a purposeful omission because the participating instructors wanted to see the results of regular work in class vs. the results of their teaching methods. Afterward, however, the changes in the different phases of the work process were measured according to the principles established by Mossop for revising texts (Mossop 2014, 134–149).

In his work, Mossop defines four overall criteria for revision of texts, which we extended in our research to apply as well to the reviewer's work. The four main criteria are the following:

1. *Problems of meaning transfer.* In our project, the meaning has to be changed to obtain the same effect of understanding among the new audience as among the source audience.
2. *Problems of content.* In our project, textual problems, for instance in the form of mistakes or lack of logic, must be corrected.
3. *Problems of language and style.* In our project, the language and style must be congruent with the style of the new audience.
4. *Problems of physical representation.* In our project, the layout, and maybe genre, of the text must be congruent with the new audience.

If we look at selected features under these four criteria, the results from the project show some interesting patterns. To illustrate point one in Mossop's list (problems of meaning transfer), we can look at the battery text example where a definition was left out but that was later required for completeness and understanding of the individual points. This was a problem of meaning transfer. For all the texts, we saw that changes were made in all three phases of the text transfer (translation, revision, review). Moreover, we saw examples of accuracy in the message being both lost and added.

Under point two (problems of content), Mossop lists logic as one important feature and factual information as another. In some cases, we saw texts in which the logic did not quite come across (e.g., the case of the introduction of wave power projects in the political system in Denmark). Sometimes lack of logic was a weak link passed on from the source text, but in some cases the review phase broke the chain of logic. This was seen a few times between the revision and the review phases; such was the case in the change of "biggest player" into "most respected player" in the example of the supplier of additives. This was both a factual mistake and an error in logic.

In general, problems of content did not appear to be a major issue. This was perhaps because the team members were quite good at asking questions of each other and giving answers. In some of the evaluations provided by students at the end of the project, a recurring comment from the US students was issues of cultural differences were not so much a matter of the language or of the examples given, but they were a factor in the topics of the texts themselves.

Problems with factual information, however, may appear in many ways, such as in the form of errors in facts or relevance of conveying facts. A clear example of that is the previously mentioned wave power text, where national politics are included. The relevance of conveying internal Danish political conditions is highly dependent on the purpose of the text and the audience who will read it. It could be argued that the political issue is irrelevant for the topic. A counter-argument is that the information would tell the audience something of the standing of the topic at a certain time and place. This situation, in turn, might or might not be replicated if the new technology were introduced into the culture of the target-language.

Problems of language and style represent a very comprehensive area in Mossop's writing. These issues include features such as smoothness, which covers the text flow and cohesion. Such factors also include tailoring, which involves formality and complexity of the text, as well as sub-language features like genre, terminology, and phraseology. This third part also includes idioms, which include, for instance, the use of collocations and rhetorical preferences, and mechanics, which includes spelling, punctuation, grammar, etc. Such basic problems represented roughly 80 per cent of all changes made, as evidenced in the many reviewing changes in the previous wave power example.

In terms of both structure and style, this example was where the native touch emanated through the texts. However, in the work process—typically in emails, but also in suggestions for change of texts—the translators and revisers were struck by the number of spelling mistakes, grammatical mistakes, and general sloppiness in the use of the language on the part of the US reviewers. The bilingual translators thought it was surprising that native speakers of English would make so many mistakes. At the same time, the bilinguals were often blind to the number of mistakes that they themselves made and their own sloppiness in email writing. Yet such factors did cause some problems for the work process, namely because the translators and revisers were very critical of the changes recommended, in part because the translators' and revisers' trust in the professionalism of the reviewers suffered badly—and vice versa.

Point four in Mossop's model (problems of physical representation) was not relevant in our projects, because the texts did not have to be formatted for publishing. This fourth point deals mainly with the layout of the text, spacing, indentations, typography, etc.

Learning from Cooperation

The experience gained by team members and project managers revealed that the translator and the reviser had to achieve consensus on the characteristics of the target audience and on the overall purpose of the text for the two parties to work fruitfully together. Moreover, guidelines had to be established to determine the extent of the revision to be made in relation to the translation. Far too often, the reviser started to retranslate the text. In such cases, the balance of revising something into a better solution or revising something into a subjectively better solution is delicate.

In the classroom, the participating instructors reached the conclusion that a line had to be drawn between "need-to-revise" and "nice-to-revise." Although this line is hard to draw definitively, a rule of thumb is that the reviser has the obligation to apply these two concepts when considering each possible revision to the text so that a change is made for a clear purpose and has a clear added value. A revision without purpose and added value cannot count as a relevant revision. This might seem a minor issue, but on a daily basis and for many texts, unnecessary changes cost too much time. Multilingual versions in T&I have to be finished *yesterday*, and revisions for the sake of revising might mean lost ground to competitors. We also found that the dialogue about one solution being better than the other is very fruitful as CCVT training—a probably highly underestimated aspect. This point has been emphasized by Schjoldager, Wølch-Rasmussen, and Thomsen (2008) who note, "Our survey found in particular that translators who carry out editing/revision need more instruction in when to correct and, in particular, when not to correct (806)."

The most important finding by the participating instructors was that the process of transition from the bilingual translator-revisor text to the monolingual reviewer text had to take place through ongoing dialogue between the team members. This dialogue was essential to ensure that no mistakes occurred during the process, because mistakes in this phase tend to fall into the category of meaning transfer, which may have severe consequences. Such consequences include the loss of conveyance of meaning and functionality and possibly the safety of a product. They also encompass confidence in the whole process from translation to review—and thus confidence in the usefulness and trustworthiness of the product.

Conclusion

The question asked at the start of the chapter was, "In trading with European nations, what do technical communicators need to know about the processes delineated in the European translation standard?" In addressing this question, the first aspect to note is the importance of the process itself. Where the focus has previously been very much text-oriented, the focus now is on the cooperation between professions and what the different professions can contribute to achieve a high-quality result. The examples analyzed showed that having complete knowledge of the different contextual situation of

- additives for the business-to-business market in two countries,
- recharging of batteries in two countries, and
- the political and technical obstacles to introducing wave power as a renewable form of energy in two countries

would surpass the competencies of a translator-reviser-reviewer operating solo.

On the whole, the projects created new insights into some of the intricacies of cross-cultural process work. They also unveiled differences that were hard to nail down but that were experienced by the students. Some of these insights call for new approaches to the process of translating-revising-reviewing as work methods. The positive outcome of the tripartite work process was that the texts in general became much more genuine in language and style and thereby also much more apt for the target-market audience. Aspects of the text that needed explicitation were identified more precisely. As a result, detailed target-market adaptations, in the form of omissions and additions, could be made more easily. In this respect, the translator-reviser-reviewer team was perfect.

The flip side of this process was that, along with the freedom of the reviewer to make the requested changes for the target market audience, misunderstandings and mistakes sneaked into the text as well. The examples discussed in previous sections of this chapter reveal the extent to which knowledge of language, semantics, pragmatics, and domain knowledge are needed to process even a fairly simple text.

It became apparent that different agents in the work process contribute different kinds of knowledge that lead to improvements in the target text. At the same time, the examples also reveal some of the pitfalls that may appear along the way. For example, the reviewer is simultaneously the strongest and the weakest point in the chain. Because the monolingual reviewer added tremendously to the quality of the text, the final texts flowed much more naturally and idiomatically, sounding as if they were composed by native speakers of English. However, when it came to the actual content of the text, the reviewers' monolingual approach was problematic. The texts moved away from the original contents, which meant the introduction of problematic aspects or even flaws. Thus, the translator's and the reviser's bilingual link between the source text and the target text helps guarantee, to a great extent, that irrelevant or erroneous information is not conveyed.

Both the positive and negative results call for added attention, especially in the review phase. The review phase did lead to improvements in the text's flow and intelligibility, but the unwarranted changes in semantics and pragmatics call for added awareness. Therefore, it is of utmost importance that dialogue be a defined and obligatory part of the process and that a focus be trained throughout on the process itself.

The education and training of technical communicators is of ample importance in this context. Without proper education and training, technical communication students will not internalize either practical knowledge or hands-on work principles, nor will they gain a gut feeling for the importance of quality assessment, quality enhancement, and value added. Further, with the growing trends of crowd-sourcing and machine translation (MT), technical communicators must become more aware of the process chain, not just the resulting text.

Notes

1 These teams are sometimes called "Globally Networked Learning Environments" (GNLEs) by scholars of international technical communication (see Stärke-Meyerring and Wilson 2008 and Melton 2008).

2 Another example is the translation of old Latin and medieval texts, which are full of implicit, time-bound clues. These texts are now usually translated by adding endless footnotes to explain context.

3 CEN members are the national standards bodies of Austria, Belgium, Bulgaria, Croatia, Cyprus, Czech Republic, Denmark, Estonia, Finland, Republic of North Macedonia, France, Germany, Greece, Hungary, Iceland, Ireland, Italy, Latvia, Lithuania, Luxembourg, Malta, Netherlands, Norway, Poland, Portugal, Romania, Slovakia, Slovenia, Spain, Sweden, Switzerland, Turkey, and the United Kingdom. In other words, all EU countries and a range of other European countries.

4 In this example, a supplier was first described as "one of the biggest," which the reviewer changed to "one of the most respected." This decision represents a good language change, but a company that is not very well known in the public can logically not be much respected.

5 The change of name from Socialdemokratiet (literally, "The Social Democracy," but a better rendering in English is "The Social Democratic Party") to Socialdemokraterne ("The Social Democrats") was changed back again to Socialdemokratiet in 2016.

References

Anawati, Danielle, and Annemieke Craig. 2006. "Behavioral Adaptation Within Cross-Cultural Virtual Teams." *IEEE Transactions on Professional Communication* 49, no. 1: 44–56.

Becher, Viktor. 2011. "Explicitation and Implicitation in Translation: A Corpus-Based Study of English-German and German-English Translations of Business Texts." Ph.D. dissertation, Universität Hamburg. http://ediss.sub.uni-hamburg.de/volltexte/2011/5321/pdf/Dissertation.pdf.

Breedveld, Hella. 2002. "Writing and Revising Processes in Professional Translation." *Across Languages and Cultures* 3, no. 1: 91–100.

Brunette, Louise, Chantal Gagnon, and Jonathan Hine. 2005. "The Grevis Project: Revise or Court Calamity." *Across Languages and Cultures* 6, no. 1: 35–36.

CEN (Comité Européen de Normalisation). 2006a. *Translation Services: Service Requirements*, 1st ed. EN 15038:2006. Brussels: European Commission for Standardization.
———. 2006b. *Oversættelsesvirksomhed: Kvalitetskrav*, 1. udgave. DS/EN 15038:2006. Brussels: European Commission for Standardization.
———. 2015. *Translation Services: Requirements for Translation Services*, 2nd ed. EN ISO 17100:2015. Brussels: European Commission for Standardization.
Chesterman, Andrew. 2016. *Memes of Translation: The Spread of Ideas in Translation Theory*. 2nd ed. Amsterdam: John Benjamins.
Cismas, Suzana Carmen. 2010. "Risk Assessment and Mitigation in Drafting Translation Contracts." *Proceedings of the International Conference of Risk Management, Assessment and Mitigation (RIMA '10)*, edited by Mircea Grigoriu, Valeri Mladenov, Cornelia Aida Bulucea, Olga Martin, and Nikos Mastorakis. 491–495. N.p.: WSEAS Press.
Ehrensberger-Dow, Maureen. 2014. "Challenges of Translation Process Research at the Workplace," ed. R. Muñoz, *Minding Translation, MonTI Special Issue* 1: 355–383.
European Parliament and Council of the European Union. 2006. Directive 2006/42/EC of the European Parliament and of the Council of 17th May 2006 on Machinery, and Amending Directive 95/16/EC (Recast). *Official Journal of the European Union*, L 157, 9.6.2006: 24–86. https://eur-lex.europa.eu/LexUriServ/LexUriServ.do?uri=OJ:L:2006:157:0024:0086:EN:PDF
European Parliament and Council of the European Union. 2017. Directive 2006/42/EG of the European Parliament and of the Council of 17th May 2006 on Machinery, and Amending Directive 2006/42/EC.
Gutt, Ernst-August. 2000. *Translation and Relevance: Cognition and Context*. Manchester/Boston: St. Jerome Publishing.
Hayes, John R. 2004. "What Triggers Revision?" In *Revision: Cognitive and instructional Processes*, edited by L. Allal, L. Chanquoy, and P. Largy, 9–20. Dordrecht, the Netherlands: Kluwer.
Herrington, TyAnna. 2008. "The Global Classroom Project: Multiple Relationships in Global Partnering." In *Designing Globally Networked Learning Environments: Visionary Partnerships, Policies, and Pedagogies*, edited by Doreen Stärke-Meyerring and Melanie Wilson, 37–51. Rotterdam: Sense Publishers.
Holz-Mänttäri, Justa. 1984. *Translatorisches Handeln: Theorie und Methode*. Suomalaisen Tiedeakatemian Toimituksia Annales Academiæ Scientiarum Fennicæ. Helsinki: Suomalainen Tiedeakatemia.
House, Juliane, and Jens Loenhoff. 2016. "Communication Studies and Translation Studies." In *Border Crossings: Translation Studies and Other Disciplines*, edited by Yves Gambier and Luc van Doorslaer, 97–116. Amsterdam/Philadelphia: John Benjamins.
Huc-Hepher, Saskia, and Barros Elsa H. 2016. "Up-Skilling through E-Collaboration." In *Employability for Languages: A Handbook*, edited by E. Corradini, K. Borthwick, and A. Gallagher-Brett, 139–148. Dublin/Voillans, France: Research-publishing.net.
Humbley, John, Bruce Maylath, Birthe Mousten, Sonia Vandepitte, and Lucy Veisblat. 2005. "Learning Localization through Trans-Atlantic Collaboration." In *IEEE International Professional Communication Conference Proceedings*, edited by George F. Hayhoe, 578–595. New York: IEEE.
Jakobsen, A. L. 2006. "Research Methods in Translation: Translog." In *Computer Keystroke Logging and Writing: Methods and Applications*, edited by E. Lindgren and K. P. H. Sullivan, 95–105. Oxford: Elsevier.

Jääskeläinen, Riitta. 1999. *Tapping the Process: An Exploratory Study of the Cognitive and Affective Factors Involved in Translating.* University of Joensuu Publications in the Humanities 22. Joensuu, Finland: University of Joensuu Press.

Katan, David. 2009. "Translation Theory and Professional Practice: A Global Survey of the Great Divide." *Hermes: Journal of Language and Communication Studies* 42: 111–153.

Maylath, Bruce, Sonia Vandepitte, and Birthe Mousten. 2008. "Growing Grassroots Partnerships: Trans-Atlantic Collaboration between American Instructions and Students of Technical Writing and European Instructors and Students of Translation." In *Designing Globally Networked Learning Environments: Visionary Partnerships, Policies, and Pedagogies*, edited by Doreen Stärke-Meyerring and Melanie Wilson, 52–66. Rotterdam: Sense Publishers.

Melton, James. 2008. "Beyond Standard English." In *Designing Globally Networked Learning Environments: Visionary Partnerships, Policies, and Pedagogies*, edited by Doreen Stärke-Meyerring and Melanie Wilson, 158–189. Rotterdam: Sense Publishers.

Metzger, Michael. 2011. "The Year of Standards? AKA Language Standards after LISA." Op-Ed. www.ata-divisions.org/LTD/documents/permanent/2011_October_Interoperability.pdf.

Milbank, Dana. 2018. "The GOP Is Buying the House. Literally." *The Washington Post*. September 26, 2018. www.washingtonpost.com/opinions/wasnt-trumps-gop-supposed-to-represent-the-little-guy-oops/2018/09/26/8a432f70-c1cd-11e8-a1f0-a4051b6ad114_story.html?utm_term=.fcbb129ae3ad

Mossop, Brian. 2014. *Revising and Editing for Translators*. 3rd ed. London/New York: Routledge.

Mousten, Birthe. 2008. "Globalisation and Localisation Influences on Web Site Text Distribution: A Case Study of Text Travel between Two VELUX Web Sites." Ph.D. dissertation, University of Copenhagen.

Reiss, Katharina, and Hans J. Vermeer. 2014. *Towards a General Theory of Translational Action: Skopos Theory Explained*. London/New York: Routledge.

Risku, Hanna. 1998. *Translatorische Kompetenz: Kognitive Grundlagen des Übersetzens als Expertentätigkeit*. Tübigen, Germany: Stauffenberg Verlag.

Robin, Edina. 2016. "The Translator as Reviser." In *The Modern Translator and Interpreter*, edited by Ildikó Horváth, 45–56. Budapest: Eötvös University Press.

Schjoldager, Anne, Kirsten Wølch-Rasmussen, and Christa Thomsen. 2008. "Précis-Writing, Revision and Editing: Piloting the European Master in Translation. *Meta* 53, no. 4: 798–813.

Schrijver, Iris. 2014. *The Translator as a Text Producer: The Effects of Writing Training on Transediting and Translation Performance*. Ph.D. dissertation, University of Antwerp.

Selby, Jenn. 2013. "David Cameron Explains Selfie with Obama and Helle Thorning-Schmidt at Mandela Memorial." *The Independent*. December 11, 2013. www.independent.co.uk/news/people/news/david-cameron-explains-selfie-with-obama-and-helle-thorningschmidt-at-mandela-memorial-8998090.html.

Sperber, Dan, and Wilson, Deirdre. 1995. *Relevance: Communication and Cognition*. 2nd ed. Malden, MA: Wiley Blackwell.

Stärke-Meyerring, Doreen, and Melanie Wilson, eds. 2008. *Designing Globally Networked Learning Environments: Visionary Partnerships, Policies, and Pedagogies*. Rotterdam: Sense Publishers.

PART III
Horizons

AREA 4

COMPUTING AND COMMERCE

8

COMPUTING AND TRANSLATION

An Overview for Technical Communicators

Tomáš Svoboda

Introduction

This chapter focuses on the theme of computers in translation[1] practices, including the advent and proliferation of machine translation (MT). The purpose of the chapter is to introduce technical communicators to central concepts that underlie translation, translation practice, and the role that translation plays in the business of global communication. The chapter first presents a general overview of how such factors developed over time. It then examines the basic terms and concepts of technology in translation. The latter includes discussions of cost, finances, and billing considerations, as well as the benefits that technology yields in communication practice. Reviewing such factors can help technical communicators better understand procedures using MT and contextualize them in relation to technical communication practices.

The chapter also provides technical communicators with practical hints to gain an understanding of translation from the perspective of content creators and/or project managers. Such information can help contextualize translation within workflow processes and production practices. The chapter then concludes with a discussion of what the future of translation technology is likely to hold. This discussion includes considerations relating to the eventually endangered status of nonprofessional and noncompetitive translators, whose work, in specific situations, might increasingly be replaced by MT technology. The examination of these topics in this way can provide technical communicators with key insights that can help guide their current and future practices as they relate to working with translation and translators.

Overview of the Topic

This section provides an overview of major computing and translation concepts, including an examination of key considerations relating to translation and computing. It also provides a review of the different processes involved to help familiarize technical communicators with such factors.

Core Concepts Associated with Computing and Translation

Today, computers and other text editing and communication devices are indispensable tools for translators/localizers and content creators alike. In many cases, messages are created and exchanged with recipients located around the globe. Yet for those messages to be effective, they need to be communicated locally: that is, in the recipient's language, rather than the sender's.

For content creators, the interconnected nature of computers and other devices and related systems offer a means to handle text material flexibly, effectively, and without needing hard copies or physical arrangements to deliver or distribute them. Such processes include a great deal of actions and subprocesses (e.g., creating, editing, and sharing texts; soliciting feedback on, inviting opinions about, and reviewing them). This lifecycle concludes with the storing, archiving, and eventually retrieving of material.

For translators and localizers, the use of computers and similar devices represents a precondition to participating in the creation process of translated/localized messages and texts. As with the content creation process, there are specialized tools for translators that include simple or sophisticated editing environments and databases of translated sentences—so-called translation memories (TM). There are also systems for delivering raw text material translated by a machine—automated translation (AT) or, as mentioned earlier, machine translation (MT), along with an array of quality-checking functionalities. In addition to these tools, there are technologies and systems to assist translators in executing an assignment, handling files, communicating with other stakeholders, planning activities, etc.

Historically, publishing practices have moved away from producing large volumes of text in one-off batches; instead, they have moved to perpetual publishing of smaller portions of text. This development has required the increased integration of translation process into the text creation process. Thus, more flexibility is required of translators to step in at whatever stage is necessary during the content creation process.

The Process: Translation as Part of the Content Management Cycle

In translation studies,[2] the notion of the translation process has been an area of research, and many scholars have attempted to define and/or visualize the translation process from various perspectives. However, one distinction is valid in all approaches:

- The translation process can involve an activity embedded in a broader communication framework: the so-called "primary communication chain," a string featuring the author of a text and its recipient. When translation comes into play, the translator steps in as a reader of the original, and the translation reaches its own readers at the end of this chain or
- translation processes can be seen as a two- or three-stage activity performed only by the translator in his/her head: the so-called "Black Box."[3]

Furthermore, due to emerging technologies in recent years, the field of specialized translation (as opposed to literary translation) has begun moving toward integration and automation. The following sections examine translation as an activity that takes place within the broader content-management process associated with such factors.

Central Concepts of the Translation Process: Aspects of Communication

To understand how translation works within content management, technical communicators must understand the evolution of thought that reflects the stages of the process in the narrower sense and content creation in the broader context. To examine such processes, this section reviews related models ranging from early concepts of content management and translation (termed here Translation Process 1.0) to the complexities of today's content management and content strategy practices.

Translation Process 1.0 (adapted from Levý 2011)

Figures 8.1 and 8.2 formalize the situations in which senders and recipients engage when communicating. The concept goes back to the early days of the discipline of translation science (i.e., the 1960s). Figure 8.1 refers to monolingual ("primary") communication and represents the sender of the original (an entity, which can differ from the text's author), and, at the receiving end, a recipient/reader/audience.

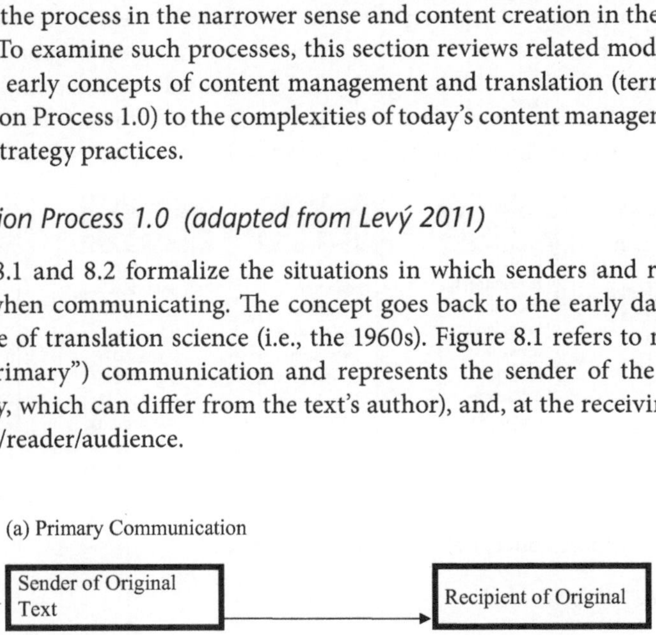

FIGURE 8.1 Primary/monolingual communication between the sender and the recipient of the original

Communication in which individuals use a language different from the original to interact with other audiences is called secondary communication. An extra communication step is inserted (the translator), yet the target group is different from that in the primary situation: Typically, the original reaches—in translation(s)—recipients not envisioned at the outset (see Figure 8.2). (In the realm of literary discourse, a book usually needs to become famous before it is translated into other languages.)

FIGURE 8.2 Secondary communication, including the translator as a special case of sender

Translation Process 1.1

When introducing other stakeholders who participate in the process, one arrives at the following chain structure (see Figure 8.3).

Some instances appear several times during the content production and translation process, namely Sender and Translation Bureau. This is because the chart reflects a chronological perspective. Here, the reader's attention should also be drawn to the two types of connecting arrows representing the original ("O") and translation ("T"). Yet this drawing does not represent all the complexities of situations when texts are conveyed to other languages and cultures. Therefore, it can still be improved to depict that the process can easily be circular rather than linear.

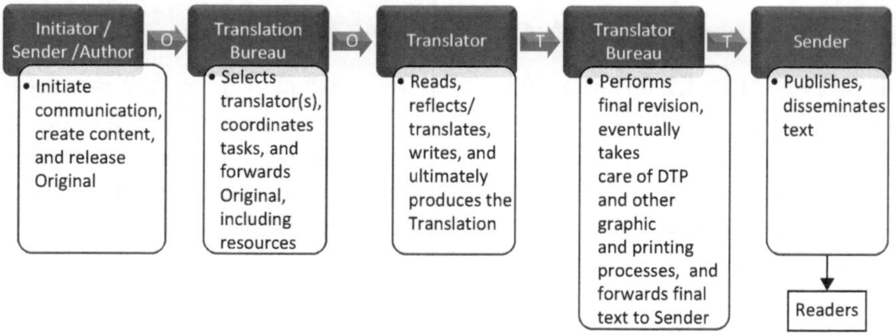

FIGURE 8.3 Linear concept of the translation process

Translation Process 1.2

In a more recent practice, workflows become much more fluid, and to capture this development, a circular chart appears in Figure 8.4.

The circular chart better captures the communication flows between stakeholders when they are no longer linear and unidirectional. The flows also

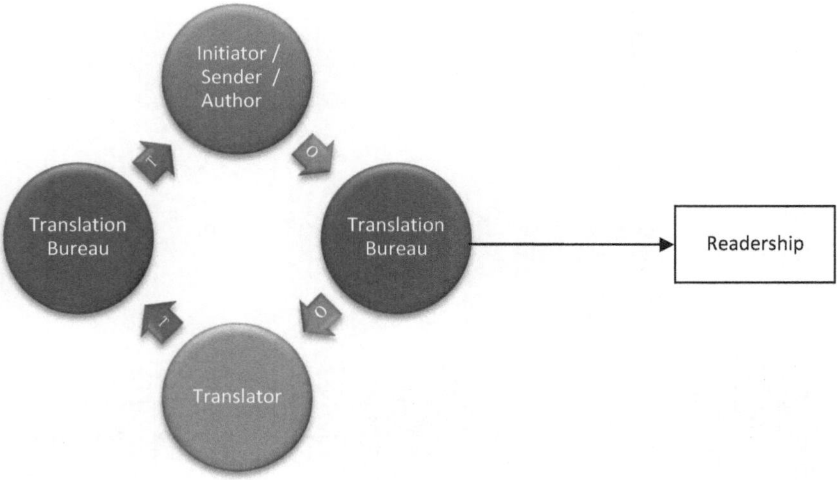

FIGURE 8.4 Circular concept of visualizing the translation process

became more fragmented and reiterated more often in the overall text creation/ translation processes. In the 1970s, for example, it was common to deliver an original manuscript to a translator and then wait a relatively long time for the translator to process the bulk text in one go (linear concept). However, advances in communication technologies (e.g., telefax) allowed for more frequent contact with translators, along with more time savings as sections of a manuscript could be delivered for translation before a complete, final version of the original was produced. The entire process speeded up, and texts, once introduced to the process, circulated in it for some time before the final product left this production platform to reach its readership.

Translation Process 2.0

In the 1990s, computing technologies started playing a role in content creation and translation processes: With (semi-)automation of text creation (i.e., authoring and smart-text editing functions), translation became inherently present in original text creation, as agents involved in content creation sought ways to make the process as efficient as possible.

Figure 8.5 illustrates how, as a result of translation automation, authoring, and process automation, original texts are created with translations in mind from the start.

In this context, authors use computer systems to screen legacy originals + translations (i.e. similar texts of the same domain/origin) to look for recyclable text material. This approach prompts authors to use a previously translated phrase (one that appears in a computer database), rather than create new ones (see Musacchio's chapter in this volume). Likewise, to save costs in subsequent/ later translation, authors are reminded not to rephrase already existing, identical sentences.

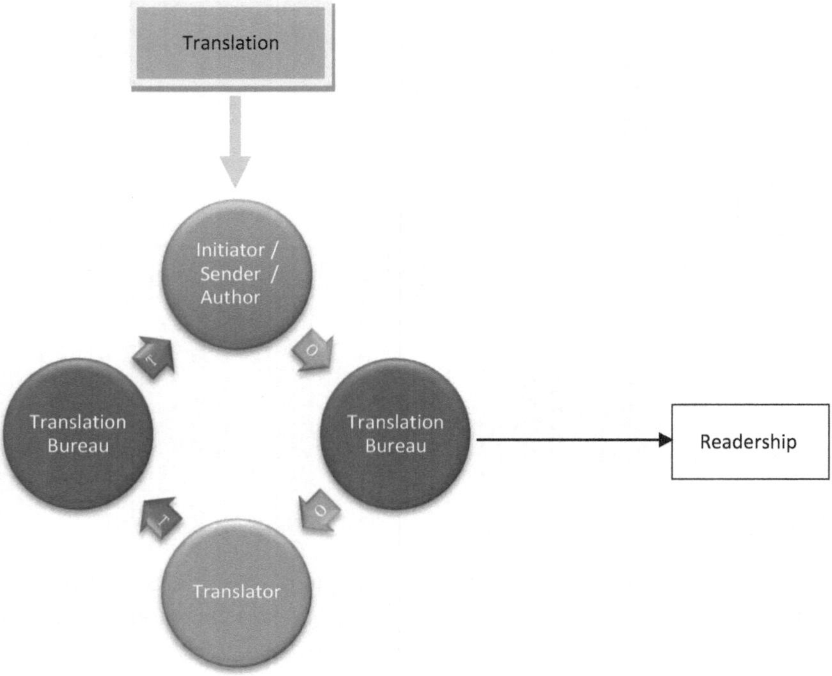

FIGURE 8.5 The idea of translation at the outset of the text-creation process

Translation Process 3.0

Quite recently, this scenario has changed in response to connectivity, ubiquitous internet access, and the miniaturization of computing devices—developments enabling rapid data exchange and mobile data communication in real time. Similarly, translators and organizations are increasingly using the so-called computer-aided translation (CAT) tools, together with other tools, such as MT, and advances in cloud computing to work more effectively. Now, instead of searching for terminology and reference documents in paper dictionaries and reference books, translators perform their searches almost exclusively online. This fact reveals that the translation process in the narrower sense (i.e., the very stage when translators translate words and sentences, apart from content creation, content management, pretreatment and posttreatment) is done almost entirely with online support.

As a result of such technological advances—especially in localization/web translation—the publishing process of a growing number of companies is fragmenting. For example, software, with its digital and searchable help manuals, is now updated/upgraded on the go and involves more minor changes done on a regular and short-term basis, in contrast to prior practices of a new releases of a revised product. Similarly, organizations tend to refrain from shipping printed technical documentation with a product; rather, they make it available either online or in an offline format that allows for eventual updating from time to time. These trends require seamless integration of translation into the publishing process. Thus, organizations can no longer view translation as external to the content-creation process. Now, companies must allow translators to log in to content management systems (CMSs) and perform partial translation assignments directly in the system as the content is being developed/created. These practices allow for the final translated versions of documentation to go live almost immediately.

As interfaces require translator input directly into content management systems (CMS), translation bureaus can be seen as obstacles in the dynamic data-exchange process. Thus, disintermediation is one of the results of the underlying integration processes. As a result, translators are increasingly using translation portals, such as Pro-Z or Translator's Café, to find clients directly (i.e., without an agent or intermediary) and to bid for assignments according to their specialization. However, bureaus and agencies will retain their place of serving as hubs for large companies requiring tailor-made and turnkey services.

Moreover, current crowdsourcing practices mean readers/users might become involved in the translation process. Similarly, the rise of online media and social networking software means senders (content creators) and recipients (content users) are much more interlinked. The whole process is thus becoming one that is interwoven, global, virtual, and instantaneous (see Figure 8.6).

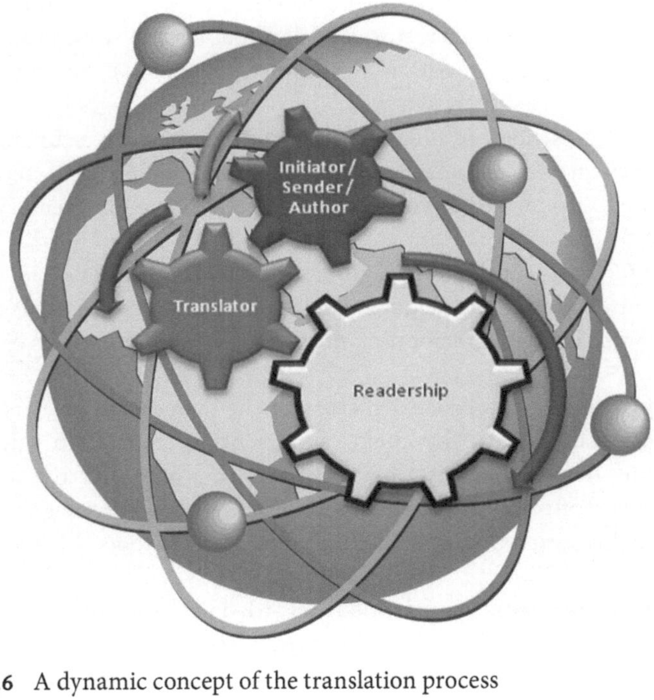

FIGURE 8.6 A dynamic concept of the translation process

In this scenario, content creators and translators work more closely together (translators are much more involved in the content creation process), while content consumers (readers) participate in content creation (e.g., microblogging)—and sometimes in translation (e.g. crowdsourcing).

The Concepts of Translation Process: Aspects of Tools

In terms of computing in general and from the point of view of translators in particular, individuals can view translation as involving six phases (Alcina 2008, 91f, abridged and updated):

- **Phase 1:** Getting a website to orient clients about working languages, specialized area(s), price per word, formats, ways of submitting source texts, infrastructure available, and so forth. Contacting other professionals by means of e-mail, mailing lists, forums, and chats.
- **Phase 2:** Sending original files as an e-mail attachment, use of virtual collaborative work environments, or downloading files by FTP.
- **Phase 3:** Documenting research by web pages. Use of general and specialized, monolingual, bilingual and multilingual electronic dictionaries on the internet (rarely on CD-ROM). Looking up information in a personal text corpus. Consulting specialists, other translators, and terminologists.
- **Phase 4:** (Translation) Using word processors. A spelling, grammar and style checker, and/or corrector. Revision tools in the word processor. Interactive assisted translation (translation memories). Use of machine translation.
- **Phase 5:** (Presentation) Preparing the translated text in a suitable format for the client. Use of e-mail to return the translated text and, should it be the case, other materials (such as glossaries, translation memory, and so forth).
- **Phase 6:** (Post-translation) Aligning original and translated texts to add them to a translation memory. Updating terminology databases. Automatic word count. Incorporating the texts in a personal corpus of texts.

We have just seen it is a broader picture of the translation process, including communication with the client or presenting translation rates on an own website, that Alcina describes. Specifically, when referring to the use of translation software, one can establish a reference between the process side and the involvement of specific functions of those tools (Melby 1998, see Table 8.1).

Here, on the one hand, translation is regarded in the narrower sense as the translator's activity (see the inner chart as opposed to the whole table—with a

TABLE 8.1 Melby's table of the use of translation technology in the translation process

INFRASTRUCTURE		
	TERM LEVEL	SEGMENT LEVEL
BEFORE TRANSLATION	• Term candidate extraction • Terminology reasearch	• New text segmentation, previous source-target text alignment, and indexing
DURING TRANSLATION	• Automatic terminology lookup	• Translation memory lookup • Machine translation
AFTER TRANSLATION	• Terminology consistency check and non-allowed terminology check	• Missing segment detection and format and grammer checks
TRANSLATION WORKFLOW AND BILLING MANAGEMENT		

three-step sequence—namely before translation, during translation, and after translation). On the other hand, the broader setting of infrastructure, workflow, and billing management is also considered as a general framework for the translator's activity. One can also view translation as creating documents/texts where various operators participate in the process.[4] To better understand the approaches involved, one needs a better understanding of terms and processes.

Computer Assisted Translation: Definition and Functions

According to (Baker and Saldanha 2009), computer-aided translation (CAT)—often used as synonymous with computer-assisted translation—means ". . . a translation *modus operandi* in which human translation (HT) is aided by computer applications. [. . .] A key characteristic of CAT is that a human translator takes control of the translation process and technology is used to facilitate, rather than replace, HT" (48).

All CAT tools come with a two-part interface that displays the original text and a space for putting in the translation. They work best with repetitive texts (i.e., originals with chunks of text, which are either similar or identically repeated within one and the same text or across several documents). Central to the system is translation memory (TM) technology (explained in more detail in the next section), which compares an original text segment against what it finds in the TM database of previously translated terms, phrases, sentences, and even paragraphs. If the database contains a matching text pair segment (original plus translation) with sufficient similarity, the tool fills the translation text into the actual translation environment. Thus, if a review of your new text finds a previously translated version of a similarly worded sentence, the system will suggest that you use that existing translation "on file," rather than translate the text from scratch.

This process leads to three scenarios: The new segment to be translated

1. exactly matches a database entry ("exact match"),
2. matches it to a certain extent ("fuzzy match"), and
3. is completely new and bears no similarity with what has been previously translated ("no-match" scenario).

Imagine, for example, a translation of a list of items in English, such as a chart of values and trends in a number of countries, as displayed in Table 8.2.

TABLE 8.2 Employees' days off

1	*Paid leave per year (in work days)*
2	Austria: 22
3	Czech Republic: 20
4	Japan: 10
5	*Paid public holidays per year (in work days)*
6	Austria: 13
7	Czech Republic: 12
8	Japan: 0
9	*Paid leave per month (in work days)*
10	. . .

This text needs to be translated into French. The translator translates the first four lines in a way that he or she types in the translation themselves. (This means that the TM does not help the translator at the outset, when it is empty.) As soon as the translator reaches the segment "Austria:" on line 6, the tool detects an exact match (the segment is the same as on line 2), because it was recently translated and automatically fills in the corresponding translation from the translation memory. The translator then reviews the proposed translation ("Autriche:") in a dedicated space for adding/editing translated text. The same is true for segments "United Kingdom:" ("Royaume-Uni:") on line 7 and "Japan:" ("Japon:") on line 8. The idea is to render as many passages of original text with translated passages on file vs. actually translating them. Still, in such cases, human involvement is indispensable, because thorough checks are necessary.

In a fuzzy match scenario, the translator translates the segment "Paid leave per year" ("Congé annuel payé") on line 1 and subsequently must translate line 9, which reads "Paid leave per month." Here, the tool detects a similar text for translation and calculates the match rate (how similar the new text is to previous text stored in a database). If this match rate is above a reasonable threshold (e.g., 55%), the tool pre-fills the translation section with text in the database/memory for the original segment (i.e., "Congé annuel payé"). Obviously (to speakers of French), this is not the correct equivalent of "Paid leave per month," so the translator has to identify the difference and edit the suggested text accordingly. In another section of the tool's interface, the difference will be highlighted to make the translator's work easier. Eventually, in the dedicated section of the tool's interface, something like this will appear:

Paid leave per ~~year~~ month Congé annuel payé

Thus, the translator can easily focus solely on the necessary amendments vs. dealing with the segment as a whole. It should be noted that the example here is a simplified description, and a range of functionalities can be set, altered, and interfered with.

In advocating for adopting and using CAT products, vendors cite increased productivity, improved consistency, better terminology handling, and enhancing traditional translation tasks—factors representing a substantial investment on the part of Language Services Providers (LSPs). Such benefits are certainly true with repetitive texts. CAT technologies should enable translators to stop wasting energy on monotonous, recurring actions and start concentrating on the actual (creative) translation process as noted in the CAT marketing literature.

Similarly, when a potential buyer considers implementing a CAT tool in her or his processes, the organization needs to consider the broader picture when contemplating return on investment and productivity gains. Pretreatment—i.e., converting non-editable formats into editable ones, the handling of TMs, the alignment of reference documents, etc.—and posttreatment—i.e., putting corrections back into TMs, post-alignment, TM maintenance, etc.—assume considerable amounts of time and resources. In terms of such topics, there seems to be a lack of reliable studies.[5] One area where types of CAT software differ is the array of supported file formats.[6]

Translation Memory

A benefit of translation memory technology[7] is that a string of text needs translating just once. From there, the translated text is entered into a database, or memory, where it is aligned to the original source text associated with it. As soon as a translated passage is stored in the translation memory database, it can be retrieved any time the source-text segment reappears in future texts. Thus, translating updates of a document or using content management systems to recycle text for new product documentation can be sped up via such technologies.

A limitation of translation memory systems is that they cannot discern between high-quality and faulty translations. During translation, once a translated segment is approved by the translator, it gets stored in the database/memory. If such a segment includes a mistake, the system will propagate the mistranslation to other places in the text where the same translation is supposed to appear. Imagine, for example, a French car manual being translated into English. The document contains the following segment: "Les conditions pour la maintenance et la réparation automobile." The first time that the translator comes across the segment, he or she translates it as follows: "The conditions for motor vehicle maintenance an repair," with a spelling mistake at the end of the sequence: instead of "maintenance AND repair" the text reads "maintenance AN repair." The translation memory system then identifies the other occurrences of the text string in other parts of the document and automatically propagates the segment to those positions. If the translator and/or reviser do not notice the mistake, it appears multiple times, instead of once. Therefore, organizations that use CAT should always store or archive only segments/translations that are finalized and correct rather than those in process.

Another consideration is the segmentation procedure necessary to construe distinct original and translation text chunks that can be paired and stored in the memory. Every language has its own segmentation rules that are predefined in the tools by default. Typically, the software will split a paragraph into sentences or segments by considering so-called start signals (e.g., capital letter) and end-signals (e.g., period/full stop). This approach is necessary to arrive at plausible segments that can be aligned or paired (original plus translation). After the text has been segmented, most CAT tools make individual segments appear on separate lines. This factor causes a potential risk on the part of the translator if she or he disregards the context of each segment and translates it inconsistently.

The pricing of CAT tools is also an ambiguous topic. With heated competition, translation buyers insist on price reductions, if originals are repetitive, as clients are not willing to pay twice for a previously translated segment that recurs in a text to be translated. To translate repetitive texts when there is an exact match in TM, it is obviously easier to double-check segments suggested by the software vs. translating them from scratch. Most CAT tools come with analysis functionalities to determine the rate of repetition (including exact and fuzzy matches, as well as not matching segments) even before starting the translation assignment. Thus, the client and the translator know in advance what level of difficulty the text will involve.

This situation is good news for translation buyers yet less so for translators. Prior to the introduction of CAT technology, translators were paid equal rates for repetitive and non-repetitive texts. Now, most often, they must agree on price reductions to stay in business. Another disadvantage for translators is the need for initial investment[8] and for initial training on the tools. Yet, due to efficiency increases that the technology makes possible, translators can compensate for the loss by producing more translated text in less time. Eventually, translators will also become more competitive when they master a tool or two required by clients.[9]

CAT Tools: The Editing Environment

The layout and basic functions of CAT tools can be demonstrated on the Man-Machine Interface (MMI) of SDL Trados Studio 2019®, a TM-based CAT tool currently considered the industry standard. The user works in an integrated environment, where many features and controls are grouped into one interface (see Figure 8.7).

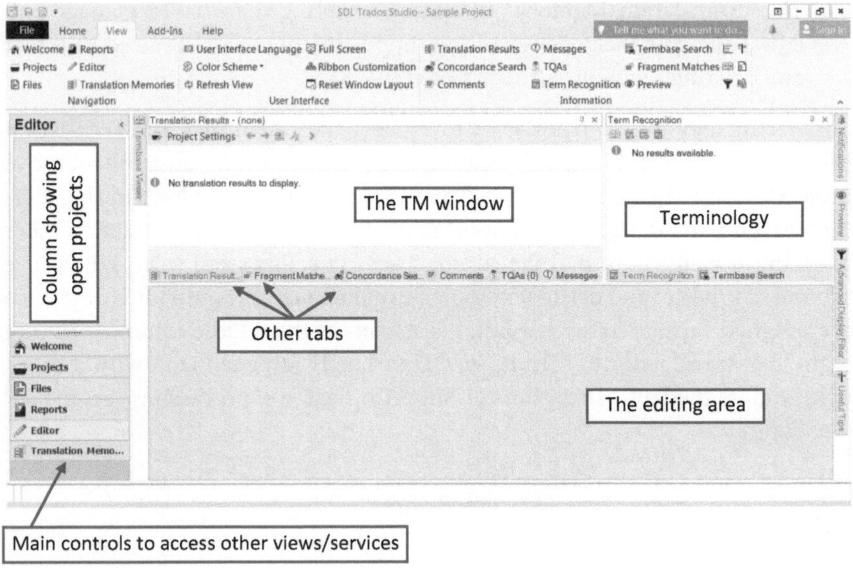

FIGURE 8.7 Trados Studio 2019®—plain interface, no translation project is open

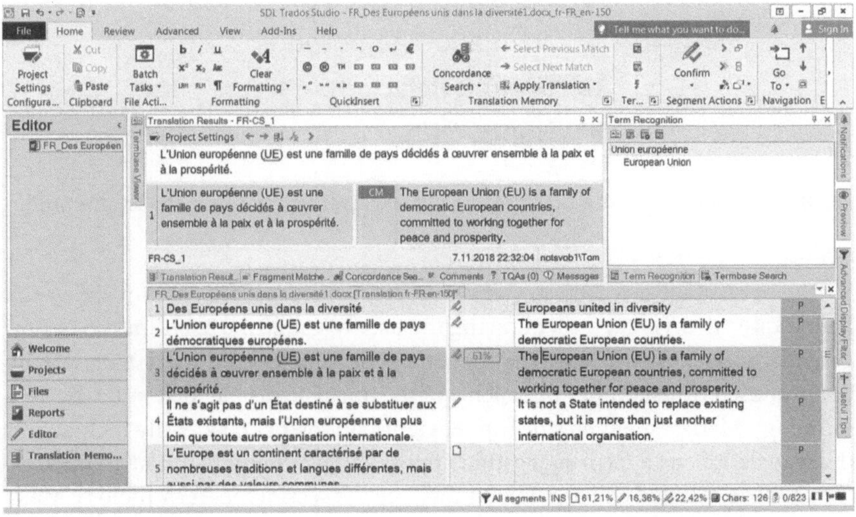

FIGURE 8.8 Trados Studio 2019®—A text open in the Editor using TM and Terminology functions

Apart from a multifaceted editing environment (see Figure 8.7 and Figure 8.8), there are other views:

- The "Home" view—a sort of a hub, where all the modules and supercontrols are bundled, equally allowing to start working on a project.
- The "Projects" interface—listing current and finished projects, allowing access to match-rate statistics.

- The "Translation Memory" interface—enabling the user to manage and edit TMs; this includes renaming, import, export as well as editing text and deleting segments already stored in TMs.

This technology allows translators to retrieve and open files and translation projects from a remote (server) location (e.g., when working on a collaborative project). While the interface's "TM" and "Terminology" windows in the "Editor" view are not designed to support text editing, the actual translation takes place in the right column of the editing area. Here, users can write, edit, move around, cut/paste, and delete text as in a usual text editor. Individual segments are assigned various status attributes (see icons in the middle column) starting with "Not translated," to "Draft," to "Translated" (a pencil icon with a green tick), and "Signed off." These statuses show the degree of processing a segment[10] (see Figure 8.9).

FIGURE 8.9 Status attributes, including corresponding icons, in SDL Trados Studio 2019®

The percentage box next to the status icon in the middle column shows the match rate vis-á-vis the TM, allowing the user to gauge the usefulness of a suggested translation proposal: The higher the percentage, the more trustworthy the proposed translation. Equivalents of technical terms blended in by the Terminology Management engine (upper right window) can be easily included in the translation, as a term suggestion appears after typing its initial characters. (For a more detailed explanation on this, see the section on Terminology Management Systems, as well as Figures 8.13 and 8.14.)

Machine Translation: Definition, Functions, Pros, and Cons

Machine translation (MT) "involves the use of computer programs to translate texts from one natural language into another automatically. It is usually subsumed under the category of computer(-based) translation, together with computer-aided translation" (Baker and Saldanha 2009, 162). Thus, machine translation, from a user's point of view, is regarded as part of a wider area where computers are used to facilitate interlingual communication.[11]

In technical communication, professional translators use MT in two ways:

- integrated in CAT tools, where MT results are displayed upon low fuzzy matches and/or no match and
- individually, in a dedicated tool that allows for editing text that had been pre-translated by an MT engine.

The latter process is referred to as Post-Editing Machine Translation (PEMT).

Post-Editing Machine Translation: The Interface

The layout and basic functions of a typical PEMT tool can be demonstrated with the help of the MMI of Google Translator Toolkit,[12] a free online tool operated by Google, Inc. (see Figure 8.10).

When an individual enters original text into GT Translator Toolkit, it is automatically processed (i.e., pre-translated) by the GT engine, and the system displays the text in the "editing mode" in the right column. With a click on a sentence, an additional window pops up (resembling a "magnifying" glass), and users can introduce editing changes to the translated text that appears in "editing mode."

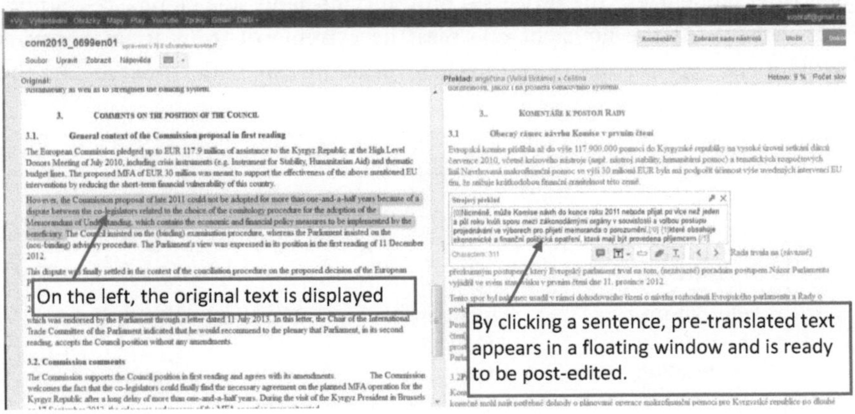

FIGURE 8.10 The MMI of Google Translator Toolkit

Generally speaking, the quality of raw MT output varies (regardless of the MT engine used), according to the language pair used, the text type, and the domain. For text types, denotative/referential texts perform better with MT compared to metaphorical/ambiguous texts. With domains, the same applies as with language pairs: The better represented a domain is in the text corpus available to the MT engine, the better results the MT system is likely to yield. For example, texts on IT in English to German translation will tend to machine-translate better than, say, Portuguese-to-Maltese translations in the area of space technology.

By offering the service free of charge, Google invites crowdsourced input by users: It records the changes introduced by users and trains its MT accordingly. Consequently, the same mistake should not reappear in automated translations in the future, as, during the so-called MT training process, the system seeks patterns in the corrective measures introduced by users and creates algorithms to prevent identical mistakes. As there are privacy concerns (the software and system are open), files/data uploaded on the GT server or in similar services should never be of personal nature. GT is thus best used for non-sensitive content and files in editable format (e.g., MS Word). It yields the best results with texts in plain, nonfigurative language, on domains that are well represented on the internet (e.g., administrative or IT) and those translated between the most frequently used world languages.

Comparison of CAT and MT

There have been several attempts to sort translation technology tools according to various criteria. One is the representation based on the level of translation automation (see Hutchins and Somers 1992; Figure 8.11). The figure shows two opposing tendencies: The processes on the right-hand side emphasize human involvement. At the far right one sees the extreme of translation scenarios without any sort of automation/mechanization involved. The processes on the

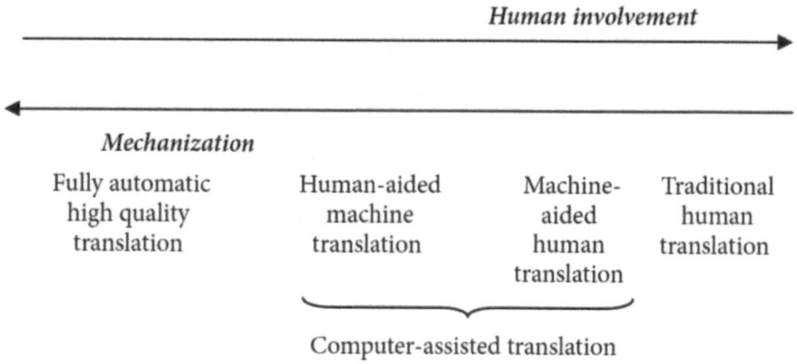

FIGURE 8.11 Hutchins and Somers (1992) classification of translation technology

left-hand side involve computer automation. At the far left one sees the extreme of fully-automated translation of a high quality. In an attempt to distinguish between MT and CAT processes, the continuum displays four translation scenarios: The two on the left pertain to MT (fully automatic high-quality translation and human-aided MT). The second from the right, machine-aided human translation, pertains to CAT. The fourth scenario, traditional human translation, at the far right, does not match either MT or CAT.

The specific feature of this division is subsuming human-aided MT (MT post-edited by humans) and machine-aided Human Translation (MAHT) under the umbrella of Computer-Assisted Translation. Here, Computer-Assisted Translation should not be confused with Computer-Aided Translation (CAT), for the former, according to Hutchins and Somers, is a concept including systems with TM technology (i.e., CAT tools) and systems in which humans post-edit machine-translated texts (i.e., PEMT environments). Equally, the term "mechanization" evokes the involvement of mechanical processes. However, "automation" would be more pertinent to current developments (work carried out in virtual realms, CAT tools running in the Cloud etc.), for neither installed software, nor virtual storage places, nor Cloud-based solutions use mechanical processes to function.[13]

A Comparison of CAT Tools

There are many differences among the dozens of today's CAT tools. They include the environment where CAT runs (installations vs. Cloud), licensing policies (virtual space rental/paying by time periods vs. fixed price per license) or the number of parameters that users can trim (e.g., to increase the tool's search performance, to customize the editing interface). When assessing such technologies, you can find a genuine comparison tool at the Pro-Z website.[14] This resource allows individuals to compare translation software, such as CAT, terminology, alignment, and others, in various categories, including price, where the choice ranges from free to US$3,500+ a year for a team edition. As of October 2018, there were 29 tools available for comparison on this site.[15]

The Special Case of Localization

The *Routledge Encyclopedia of Translation Studies* (Baker and Saldanha 2009) gives the following definition for localization:

> Localization can be defined as the linguistic and cultural adaptation of digital content to the requirements and locale of a foreign market, and the provision of services and technologies for the management of multilingualism across the digital global information flow.
>
> *(157)*

In practical terms, localization (commonly abbreviated to "L10n" due to the 10 letters between the initial "L" and the final "n" in the word) is a trendy expression for standard translation performed with some restrictions or under special conditions. The former includes text-length restrictions when translating/localizing software. The latter includes playability considerations when localizing PC games.

An example of text string restraints is shown in Figure 8.12. Here, software localizers and checkers of a web browser program did not spot a mistake in the translation equivalent used as shown in the highlighted button. Its final letters and an additional word are missing.

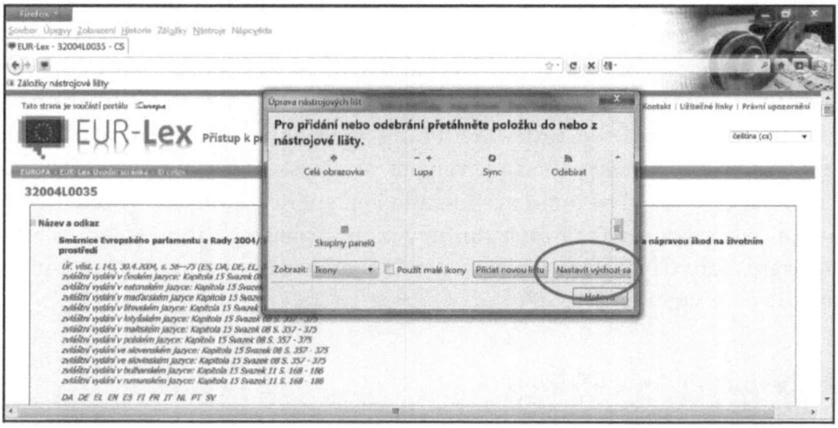

FIGURE 8.12 Example of an error when localizing dialog boxes

Instead of "Nastavit výchozí sa" ("Define a default set of bu"), the button text on the right should read "Nastavit výchozí sadu tlačítek" ("Define a default set of buttons").

When the source culture that produced the original text and target cultures for which the text is to be translated are not very different, all that really needs to be done is linguistic adaptation (i.e., converting the text from one language into another) vs. repurposing non-linguistic elements such as design, layout, image use, etc. Such text-only processes are part of professional translation practice anyway. When the cultures differ substantially, however, the localization (i.e., complete redesign of features and formats beyond the text alone) of websites may include adapting the page layout, color background, website music, graphics, and the use of pictures.

On the textual plane, there might be differences in the level of information detail, text-sorting conventions, listing rules, etc. Particularly interesting is the issue of search engine optimization (SEO), which affects the visibility of a website within a search result list. Here, translators/localizers must adhere to specific requirements regarding the frequency and location of keywords in the translation to make the translated (localized) site as search-engine friendly as the original.

Terminology Management Systems

Technical translation requires harmonized terminology: using one specific term per one notion and using it consistently throughout a content creation/translation project. Terminology management is a way of handling recurring and established terms to store, retrieve, and edit them (both the original term and its equivalent) in a databank. Accordingly, numerous professional organizations, including FIT (Fédération Internationale des Traducteurs, the world association of professional translators) acknowledge the profession of terminologist. A terminologist's job involves determining harmonized usage of defined/fixed terms throughout a text (typically a translated text). This includes term extraction (looking for terms in the original text), research for equivalents, and ensuring that one term is rendered in a unified way throughout a translation project. Whereas full-time terminologists are hired only in large companies or institutions with a sophisticated language management strategy, terminology work is carried out on an everyday basis by practicing translators and interpreters as well.

There are software tools to help translators achieve the kind of harmonization mentioned here, and information and communications technology (ICT) helps with key tasks associated with such processes, including terminology extraction, sorting, storage, lookup, putting in translation, and checks. Most notably, today's integrated CAT tools are shipped with a terminology management component: a database, where terms can be stored and subsequently retrieved in the translation. This factor allows CAT tools to draw from databases of previously compiled, translated text when translating new documents. When the terminology recognition function is turned on in a CAT tool, terms that had been defined in the associated term-management component (i.e., a database of previously translated terms) are shown in a separate window (see Figures 8.7 and 8.8). This means that the user receives an initial translation suggestion, instead of having to look up term equivalents elsewhere, and over and over again. Figure 8.13 provides an example of the practical usage of a defined and stored term.

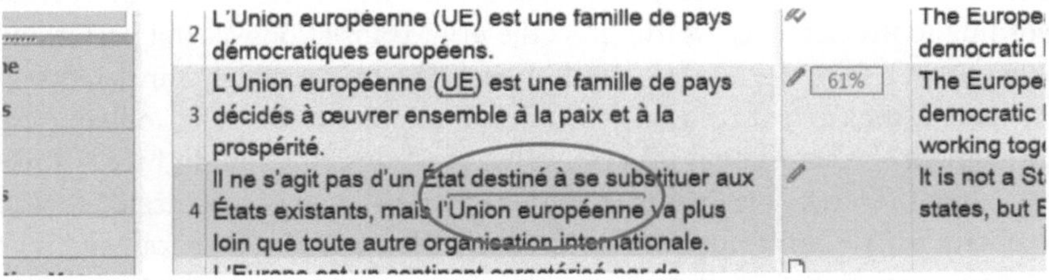

FIGURE 8.13 Terminology management in SDL Trados Studio 2019®: Term recognition

Here, the translator previously stored a term in the tools' terminology database with "Union européenne" being the source term. (In Figure 8.13, see the horizontal line above the string of characters "Union européenne" in the left-hand side column, i.e., the original). The English equivalent reads "European Union." In Figure 8.14 in the right-hand column, note that the translator just started to write the first two letters of the equivalent ("Eu . . .") and the context menu appears with the term prompt, which can easily be accepted by pressing "Enter." In general, the availability of online technical resources and the advancement of corpus linguistics have benefitted terminology management immensely.

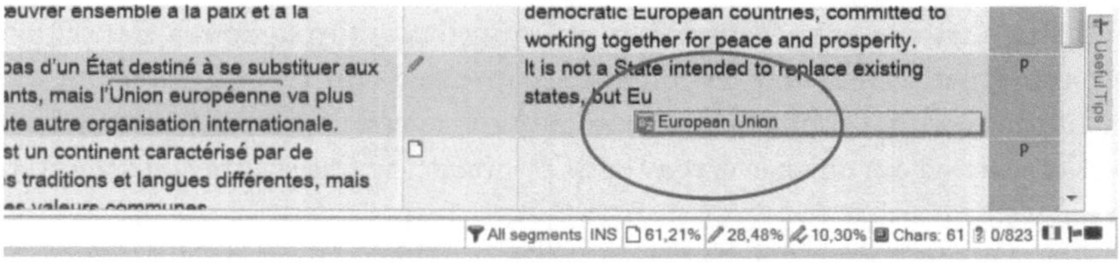

FIGURE 8.14 Terminology management in SDL Trados Studio 2019®: Term equivalent as recalled from database and offered for insertion

Text Alignment Tools

Text alignment is the process of linking segments of original text with those of the text's translation. It is used to create a database of matched segments for eventual use in a subsequent assignment. Imagine an annual company report translated last year without CAT tools. This year the new annual report is quite similar: just a few minor parts have changed—namely, the introduction and a few pages on innovations. Overall, text sequences on reporting topics need to be kept the same; only the figures have been updated. Thus, what we have is a new translation assignment here, with the client having provided last year's annual report as a binding reference. For the translator, it would be an asset if she or he had the previous version stored in the TM to see immediately in a CAT tool the level of text segment matching when the tool compares the current wording to the reference wording. As the prior translation was not performed using a CAT tool, there is no TM featuring the original and translated segments that the current translator could feed into her or his CAT tool and use throughout the translation process. What the translator can do here is link/pair/align the legacy original with the legacy translation in a dedicated tool so to obtain a TM. This information can then be fed into the CAT tool and thus be made operational when dealing with identical text segments in the current assignment.

To summarize, two types of situations can occur:

1. Situations where prior translations were not performed with the help of a CAT tool
   a) Reference documents (original plus translation) show, on segment level, considerable similarity with an original currently submitted for translation.
   b) Reference documents exist that are imperative to refer to during a current translation assignment; the typical reference level is a phrase or a term.
2. "Post-alignment," a cumbersome procedure of aligning finalized translations with originals after completing revision/proofreading/polishing of the translations. Because the text-improvement tasks of revision, checkup, and polishing are typically performed outside CAT tools, amendments need to be put back into the TM used for translating. Today there are ways of exporting the translation text from a CAT tool, making it available to specialists (who might not be fully conversant with translation tools) for revision/proofreading and then importing the polished text back into the CAT tool. Thus, what is stored in the CAT tool in the end is the finalized wording of the translated text. This approach makes the TM a reliable, quality resource for similar translation assignments.

Project Management Systems: Pros and Cons and Functions

In translation practice, the project management (PM) functionality of a system is incorporated in integrated interfaces or is performed externally in a dedicated interface. PM applications include the following:

- a time management function (calendar, individual project/assignment tracking, time/progress indicators)
- sometimes, a monitoring channel for project managers in the translation agency to follow work progress
- collaborative features
- utilities to pack/unpack and send/receive project packages (to include TM, glossaries, and reference documents)

Graphic features, such as progress bars or alerts, and communication or notification facilities may also be included.

Costs, Finances, Billing, and the Factors Affecting Them

The introduction of CAT tools saw a major change in translation billing practices. From the outset, translation services providers have considered the

TABLE 8.3 Example of pricing vis-à-vis fuzzy percentage bands and segment status

Category	*Description*	*Percentage band*	*Pricing*
100% match	For respective source segments, there are exact translation matches in the memory.	–	25% of full price
Fuzzy match	For respective source segments, there are similar, but not exact, translation matches in the memory.	75% to 99% similarity	75% of full price
Zero or low fuzzy match	There are either no or similar (yet irrelevant) translation matches in the memory.	0% to 74% similarity	full price
Repetitions	Segments in this category are repetitions of the same texts strings either within the same translation file or across several files in a project. They are pre-filled automatically by the tool.	–	25% of full price

technology as a way to reduce human translation costs by paying less for repetitive text and reducing the need to repeatedly translate the same text. Over time, certain percentage bands have developed that mirror the effort needed to edit repeating text (see Table 8.3).[16] Individual translation service providers or translators may apply varying pricing methods.

According to this perspective, some price is paid even for exactly matching or repeating segments. This is substantiated in the way that even those translation text segments (matches) that come from the memory and are inserted automatically by the tool (i.e., with zero effort by the translator) need to be checked for consistency and fitness for the given context. So, whereas the CAT tool takes care of searching the TM database and inserting the most likely equivalent, the translator is responsible for checking the translations and considering their appropriateness. Rates for repeated and exactly matching segments should be similar to rates charged for a revision job.

In terms of MT pricing, one always pays for machine translation. One or the other of the following is always in play:

- There are hidden manpower costs behind the technology because an LSP who intends to introduce MT needs to hire a computer linguist for the technology's introduction and maintenance.
- Engines free of charge for the public are not always free to use under commercial terms. For example, Google Translate (GT) is virtually free when accessed on the Google web page (users have to subscribe and by using the service their human editing input is used to train the MT tool), but GT will charge users US$20 per one million characters of text if they have it turned on within their CAT tool.

For completeness, it should be noted that there currently are free MT engines available in the form of an API for CAT tools. This scenario considers MT to run within a translation (CAT) environment with no implications on translators' word rates. However, price reductions are applied when post-editors edit the output of a proprietary MT of sufficient quality (i.e., in dedicated tool environments). In such a case, there seems to be no best pricing practice yet. However, because MT output can vary depending on language pair and domain, the recommendation is to charge PEMT services by the hour.

The Benefits of the Tools and Technologies

As explained in the subsection titled *The Concepts of Translation Process: Aspects of Communication*, language technologies have become an integral part of market practice today because they offer flexibility in drafting, editing, reviewing, and finalizing texts. The market of language-/text-processing technology is becoming increasingly complex, and it develops rapidly, resulting in dynamism requiring a flexible response on the side of LSPs. To illustrate the wide concept of today's translation market, one can confront various adjectives used to describe its features, as one sees in Olvera-Lobo et al. (2005, 132–135):

- the global market
- a decentralized market
- a specialized market
- a dynamic market
- a virtual market
- a demanding market

Additionally, the services and products involved in translating are becoming increasingly virtual (the rate of text published online is growing steadily), texts are consumed more rapidly, they tend to be short-lived, and thus they need to be replaced by new content soon. These developments equate to shorter turnaround times when language versions of texts need to be produced. Productivity gains and fierce competition pressures are reflected in reduced prices. At the same time, speed and price constraints call for further increases in effectiveness, with some LSPs introducing MT engines to produce rough translations later refined by post-editors.[17]

As one might imagine, this context means quality considerations are substantiated when it comes to translation. As PEMT jobs are assigned in projects favoring short deadlines over quality, post editors are expected to concentrate on meaning and disregard language register and style. As in other sectors, in translation the saying holds true as well that you get what you pay for: Paying less might save financial resources, yet unsatisfactory quality output can result in much higher volumes of human resources wasted by having to correct poorly translated texts.

The Process: Connection to Professional Communication Practices

The following recommendations apply in streamlining content creation, translation, localization, and/or internationalization processes.

Translations—Doing Them Right: Preliminary Thoughts

To begin, sender of a text should ask if translations are necessary at all. If so, language pair(s) (i.e., what language versions need to be produced) and specific volumes (i.e., what portions of a source text need to be rendered) should be specified. Demand management—the control over amounts of text assigned for translation—is a key concept under budgetary constraints. Translation buyers should always keep in mind that high-quality translation is an expert activity requiring appropriate funding. The whole communication endeavor needs to be tailored according to the expectations of the potential readership.

To address such factors, clients must communicate the purpose of a source text and a translated text to the translator. Such communication is essential, for there is a substantial difference between translating a text for informative purposes only and one that is destined for publication in high-profile publications. Moreover, the mother-tongue principle (i.e., translators translate only into their mother tongues/native languages, not the reverse) should not be compromised. Additionally, to be complete, professional translation assignments must be revised by a second professional translator.[18]

Editable Originals

Currently, a growing proportion of text material identified for translation is disseminated in non-editable formats (e.g., pdf files). As this factor may stem from the desire to respect intellectual property, try to avoid sending non-editable file formats for translation. Otherwise, translators will spend a considerable amount of time converting the text into an editing format (typically with Optical Character Recognition/OCR software) and may spend less effort producing top-quality translations. Fortunately, sophisticated CAT tools allow editing of the text layer in native file formats produced in DTP applications, such as QuarkXPress, Adobe FrameMaker/PageMaker/InDesign, AutoCAD, and others. Thus, there is no need to transfer text material from those tools into a Word file or during a cumbersome reimport after translation.

Authoring: The Effective Way to Produce Source-Text Materials

When one is judging the quality of translation output, both CAT and MT technologies depend on the quality of source texts. Whereas what counts with CAT is structural similarity or identity (such as similarities in punctuation and sentence structure in the source vs. the target texts), MT engines may require source texts to be drafted according to the requirements of controlled language (see O'Brien's chapter, this volume). Such requirements include avoiding overly long sentences, using sentences with a restricted number of subordinate clauses, limiting the modality of verbs, reducing the richness of vocabulary, and using explicit grammatical structures.

Style Guides: The Keys to Success

Typically, translation buyers initiate the creation of an authoring and/or translation style guide. To avoid trial-and-error processes, language service departments can begin developing such style guides proactively. These may include chapters on general requirements (e.g., completeness, file formats, file handling, deadlines), translation procedures, terminology, style, grammar, and more. Every major institution or multilingual company will likely have a style guide. The European Commission's Translation Services publishes its style guides online at http://ec.europa.eu/translation/english/guidelines/en_guidelines_en.htm.

Consistency Across Texts Is Best Achieved by a TM and Terminology Management

In conservative domains, such as professions associated with law and the production of legal documents, where unambiguity is a decisive factor, consistency in the use of phrases and terminology is important. Translators can use TM software and terminology management tools to achieve such consistency. The sooner the tools are implemented, the less effort is needed for searching, referencing, and text-alignment to achieve the goal later on.

Projections for the Future

The production of content—that is, texts to be communicated—is predicted to increase (see www.taus.net), and with exponential growth, the only way to make it accessible in other languages in its entirety is said to be MT (depending on language combination and domain, of course). While CAT technology represents an enhancement of word processing software, MT is the paradigmatic change in the profession and industry. The developments ahead could hardly be more profound in the near future.

Today, we are witnessing the massive proliferation of MT, as it becomes a standard feature of a number of websites and is shipped with some internet browsers free of charge.

The initiative has shifted from translators to computational linguists. In the area of machine translation, what used to be state-of-the-art MT just a few years ago (statistical MT, also known as SMT), is being replaced by neural MT

(NMT) today, arguably yielding still better results. MT service providers who manage to receive feedback from MT users on the MT output—for example, by collecting crowdsourced post-edits—use it to fine-tune and further train and improve their MT engines.

In the short-term, the future will witness the proliferation of proprietary MT. NMT engines will eventually be complemented by rule-based MT (RBMT) and specific language corpora. Offline MT is the challenge of today. The quest for high-quality MT will continue with MT engines being trained to cover specialized domains.

In terms of the market, disintermediation (a stable pool of translators working directly for a client, while eliminating the mediation costs charged by translation bureaus) and online-ization (collaboration in virtual realms) will prevail, leaving ineffective translation agencies behind. The market will continue to embrace social media features, such as the rise of collaborative platforms with peer translator and/or client rankings, real-time chat-like translation services, and will see artificial intelligence (AI) kick in with translation quality estimations and with translators automatically ranked based on a number of aspects, such as timeliness, product quality, positive/negative feedback, availability, etc. MT will reduce jobs for nonprofessional translators, and human translation will focus on highly specialized domains and creative texts. In non-specialized domains, turnaround times and prices are bound to keep dropping for some time, forcing noncompetitive actors out of the market. After that time, in the long-term perspective, high quality human translation will become a specialized, respected, and luxury kind of service.[19]

Conclusion

The translation market and the translation and localization profession are witnessing profound changes due to the introduction of language technology solutions, most notably MT. To provide translation its rightful position within the communication process, content creators and project managers can use the opportunities that technology offers to streamline underlying processes. One central suggestion is to keep translations in mind as soon as planning a communication project and creating original/source texts begins.

Understanding the distinctions among individual tools used in providing language services is crucial in deploying them in the most effective manner. Inappropriate use of these tools, or the failure to use them altogether, leads to losses in effectiveness, while successfully implementing them boosts service quality and job satisfaction while resulting in effective communication.

Machine translation will likely reduce the job prospects of nonprofessional translators. It may also interfere with language development and result in reduced language richness and diversity. On the other hand, it has already become part of online communication practices. Language professionals should not despair. When responsibly implemented, embraced, and channeled in an informed manner, MT can turn from “bad master” to a “good servant.” Translation and localization are bound to remain vital in global communication practices.

Notes

1 Unless specified otherwise, translation refers to the area of non-literary translation here.
2 The discipline of translation studies (also referred to as translatology) was established in the 1960s and early 1970s. Today, it is interdisciplinary, encompassing (written) translation, (oral) interpreting as well as localization, and terminology work.
3 A three-stage model would, for example, include reading the original text in the source language, interpreting its message and reformulating the message in the target language.
4 For a discussion of some other classifications, see for instance Král 2012.
5 As Jiménez-Crespo (2009) notes, “Nevertheless, even when it has been previously suggested that TM tools bring about increased quality and consistency . . . , there is a scarcity of product-based empirical studies that compare texts translated using TM tools with those produced without them in order to validate some of these underlying assumptions” (215).
6 By way of example, a list of file formats supported by the SDL Trados Studio 2019, the leading CAT software, is presented in Appendix 8.1.
7 TM is a crucial part of CAT tools. For an introductory discussion on TM, see the preceding section.
8 Please note that pricing policies among CAT tools vendors differ and that today there are tools available for free or with a time-bound licensing policy rather than invoking a charge for installations.
9 For further considerations on the limitations of CAT tools, see Mačura 2012.
10 There are seven statuses to choose from. A translation process in the Editor is finished after all segments have the status “Translated” or “Signed off,” depending on whether a revision step is included in the process.
11 Unfortunately, little comparable data is available to determine what texts are most frequently rendered/translated through freely accessible MT engines. According to Seljan (2011, 283), roughly a third of such texts are “professional texts.”
12 Not to be confused with Google Translate. Whereas the latter is a free online service to receive instant translations in a very simplified interface, the Toolkit is accessible only for registered users. It still is free to use and has advanced features, such as the option of uploading translation memories, as well as more sophisticated editing functionalities.
13 A comparison of statements/assessments as regards both the CAT tools and machine translation is given in Appendix 8.2. It has been elaborated based on available literature and the author’s own input, with aspects grouped into four thematic groups: the systemic level, the process, the product, and the user’s view.
14 Available from: www.proz.com/software-comparison-tool (retrieved 24 October 2018); Pro-Z is the world’s largest translator directory and freelance translation portal.
15 A much more reduced number of comparison categories can be found at the Wikipedia page on CAT (available from: http://en.wikipedia.org/wiki/Computer-assisted_translation, retrieved 24 October 2018). Some of the Website’s language versions include the price feature in the comparison as well.

16 NB: The suggested prices and fuzzy match bands are presented by way of example here.
17 PEMT is said to yield as much as 30% to 80% productivity gains on the part of the translator (see Král 2012, 193).
18 For further information, Chris Durban's (2011) mini-guide is a valuable source.
19 For further considerations on future development of the translation profession, see Palumbo's chapter, this volume, and Svoboda 2017.

References

Alcina, Amparo. 2008. Translation Technologies: Scope, Tools and Resources. *Target: International Journal of Translation Studies* 20, no. 1: 79–102.

Baker, Mona, and Gabriela Saldanha. 2009. *Routledge Encyclopedia of Translation Studies*. 2nd ed. London: Routledge.

Durban, Chris. 2011. *Translation: Getting It Right*. Alexandria, VA: American Translators Association.

Hutchins, W. John, and Harold L. Somers. 1992. *An Introduction to Machine Translation*. London: Academic Press.

Jiménez-Crespo, Miguel A. 2009. "The Effect of Translation Memory Tools in Translated Web Texts: Evidence from a Comparative Product-Based Study." In *Evaluation of Translation Technology* 8, edited by Walter Daelemans and Véronique Hoste, 213–232. Antwerp: Artesis Hogeschool Antwerpen, Departement Vertalers en Tolken.

Král, Pavel. 2012. "The Role of Technology in Translation Studies." In *Teaching Translation and Interpreting Skills in the 21st Century: Proceedings of the International Conference Translation and Interpreting Forum Olomouc 2011, Organized by Department of English and American Studies, Philosophical Faculty, Palacký University in Olomouc, Czech Republic, November 11–12, 2011: TIFO 2011*, 189–205. Olomouc, Czechia: Palacký University.

Levý, Jiří. 2011. *The Art of Translation*. Translated by Patrick Corness. Amsterdam/Philadelphia: John Benjamins.

Mačura, M. 2012. "Limitations of CAT and MT Technology." In *Teaching Translation and Interpreting Skills in the 21st Century: Proceedings of the International Conference Translation and Interpreting Forum Olomouc 2011, Organized by Department of English and American Studies, Philosophical Faculty, Palacký University in Olomouc, Czech Republic, November 11–12, 2011: TIFO 2011*, 207–219. Olomouc, Czechia: Palacký University.

Melby, Alan K. 1998. "Eight Types of Translation Technology." Paper presented at the 39th Annual Conference of the American Translators Association (ATA), Hilton Head Island, SC, November 4–9.

Olvera-Lobo, Maria Dolores, Maria Rosa Castro-Prieto, Enrique Quero-Gervilla, Ricardo Muñoz-Martín, Eva Muñoz-Raya, Miguel Murillo-Melero, Bryan Robinson, José Antonio Senso-Ruiz, Benjamín Vargas-Quesada, and Cristina Domínguez López. 2005. "Translator Training and Modern Market Demands." *Perspectives: Studies in Translatology* 13, no. 2: 132–142.

Seljan, Sanja. 2011. "Translation Technology as Challenge in Education and Business." *Informatol* 44, no. 4: 279–286.

Svoboda, Tomáš. 2017. "No Linguistic Borders Ahead? Looking Beyond the Knocked-Down Language Barrier." *TranscUlturAl: A Journal of Translation and Cultural Studies* 9, no. 2, 86–108.

APPENDIX 8.1

FILE FORMATS SUPPORTED BY SDL TRADOS STUDIO 2019

SDL XLIFF Documents (*.sdlxliff)
TRADOStag Documents (*.ttx)
SDL Edit Documents (*.itd)
Microsoft Word 97-2003 Documents (*.doc;*.dot)
Microsoft Word 2000-2003 Documents (*.doc;*.dot)
Microsoft Word 2007-2016 Documents (*.docx;*.dotx;*.docm;*.dotm)
Microsoft PowerPoint 97-2003 Documents (*.ppt;*.pps;*.pot)
Microsoft PowerPoint XP-2003 Documents (*.ppt;*.pps;*.pot)
Microsoft PowerPoint 2007-2016 Documents (*.pptx;*.ppsx;*.potx;*.pptm;*.potm;*.pps
Microsoft PowerPoint 2007-2016 Documents (*.pptx;*.ppsx;*.potx;*.pptm;*.potm;*.pps
Microsoft Excel 2007-2016 Documents (*.xlsx;*.xlsm;*.xltx;*.xltm)
Microsoft Excel 2007-2016 Documents (*.xlsx;*.xltx;*.xlsm)
Microsoft Excel 2000-2003 Documents (*.xls;*.xlt)
Bilingual Excel Documents (*.xlsx)
SDL Trados Translator's Workbench Documents (*.doc;*.docx)
Rich Text Format (RTF) Documents (*.rtf)
Microsoft Visio Documents (*.vsdx;*.vsx;*.vtx;*.vdx;*.vssx;*.vstx;*.vsdm;*.vssm;*.vstm;)
XHTML 1.1 Documents (*.html;*.htm)
Html Documents (*.htm;*.html;*.xhtml;*.jsp;*.asp;*.aspx;*.ascx;*.inc;*.php;*.hhk;*.hhc)
Html Documents (*.htm;*.html;*.xhtml;*.jsp;*.asp;*.aspx;*.ascx;*.inc;*.php;*.hhk;*.hhc)
Adobe FrameMaker Documents (*.mif)
Adobe InDesign INX Documents (*.inx)
IDML Documents (*.idml)
Adobe InCopy CS4-CC ICML (*.icml)
OpenDocument Text Documents (*.odt;*.ott;*.odm)
OpenDocument Presentation Documents (*.odp;*.otp)
OpenDocument Spreadsheet Documents (*.ods;*.ots)
QuarkXPress Export Documents (*.xtg;*.tag)
XLIFF Documents (*.xlf;*.xliff;*.xlz)

XLIFF: Kilgray MemoQ Documents (*.mqxlf;*.mqxliff;*.mqxlz)
XLIFF 2.0 Documents (*.xlf;*.xliff)
WsXliff Documents (*.xlf)
PDF Documents (*.pdf)
Comma Delimited (CSV) Documents (*.csv)
Tab Delimited Text Documents (*.txt)
Java Resource Documents (*.properties)
Portable Object Documents (*.po)
SubRip Documents (*.srt)
JSON documents (*.json)
Microsoft .NET Resource Files (*.resx)
DITA XML Documents (*.xml;*.dita;*.ditamap)
DocBook XML Documents (*.xml)
Author-it XML Documents (*.xml)
MadCap XML Documents (*.html;*.htm)
ITS XML Documents (*.xml;*.its)
XML Documents (*.xml)
Text Documents (*.txt)

APPENDIX 8.2

COMPARING CAT AND MT (PEMT)

The Systemic Level

TABLE 8.4 Comparing CAT and MT (PEMT) tools on the systemic level

	CAT	*MT / PEMT*
History	Beginning and spread in the 1990s.	Beginnings in the 1950s. Around 2008 start of widespread interest in the translation community; massive proliferation since then.
Development	Mature system, reaching its boundaries.	Dynamically developing, with challenges such as the "Time Loop"—a situation where MT engines start using text material produced by (other) MT systems, thus errors get recycled; also known in comparable terms as "Point of MT Degradation."
Future	Less important as a stand-alone solution, but still valid in connection with MT implementation.	Big demand.
Paradigmatic	No; not disruptive. The technology does not "translate" instead of the translator; it just reduces redundancy by providing the translator with already translated text chunks where appropriate.	Yes; the translator does not translate from scratch any more. He/she edits text material that has already been rendered by the machine.

(*Continued*)

	CAT	MT / PEMT
Proliferation	Professionals; people outside the translation profession hardly ever hear of CAT tools and do not use them.	Ubiquitous; publicly available MT engines and interfaces are an integral part of the search experience and some online platforms/web pages.
Technology	One; TM—a database storing both original and translated text segments.	Various (EBMT, RBMT, SMT, NMT, hybrid/mixed . . .).
Tertium Comparationis	None: Source language segments are paired with Target language ones directly without a mediating (meta) language.	Yes with some publicly available MT engines, such as Google Translate. No with some customized/ proprietary tools (especially RBMT).
Supported language pairs	Democratic: It makes no difference if languages with larger or smaller diffusion are used.	Uneven/Undemocratic: The dominant vehicular language in most MT engines is English (i.e. translations running from languages other than EN are rendered via English).
Data retrieval	Offline/Online; can work using TM from a remote location (e.g. server) as well as locally on the hard drive.	Offline/Online: The most typical use is online (the system is dynamic and variations can be observed often). However, there are offline MT solutions available working with data sets stored locally (e.g. downloading an MT solution for a specific language pair).
Cost	Various pricing methods, from free to high cost (for agency and for translator).	Various pricing methods, from free to high cost (tailor-made corporate MT architectures).
Confidentiality	Little difficulty: texts can be kept secret in in-house, offline, or firewall protected server-based scenarios.	Complex issue: Unless the MT engine is run within a corporate firewall, the use of publicly available MT engines is always a threat to confidentiality of texts processed by the engine.
Speed	Almost instant response (although there are time-consuming tasks, yet outside the actual interactive translation process, such as Converting to bilingual file, Import/ Export, Re-Organizing TMs).	Almost instant response when working in a web-based interface (with broadband internet connection). However, when using MT to translate entire texts in one go, response time can become rather extensive (probably also depending on server traffic and computation load). When working in a CAT-tool interface and using an MT plug-in (i.e. the "Interactive" mode), even with high-speed internet connection, there can be longer response times resulting in user dissatisfaction.

	CAT	*MT / PEMT*
Usability	Restricted, narrow: Best results are achieved on certain text types/ domains only (e.g. formalized language, administrative style, forms, lists). Mainly suitable for repetitive texts, since with texts of low repetitiveness rate (i.e. low match scenarios) TM technology provides minimum assistance.	Restricted, yet much wider: Although best results are achieved on certain text types, domains only (e.g. formalized language, administrative style), repetitiveness is no precondition. For creative language, literary works, the amount of post-editing effort usually makes the use of MT prohibitive.
Customizability	Partial (match rate threshold, penalties).	None or little (publicly available MT engines); Yes (custom/proprietary engines).
Upfront investment	Yes, but lower than for proprietary MT.	Lower or none with publicly available MT engines (suitable for individual translators). Considerable investment necessary in proprietary MT for large LSPs.
Maintenance	Yes: It may be necessary to perform TM maintenance tasks from time to time. TMs have to be sorted, stored, exported, imported, filtered, etc.	No: Publicly available MT engines come maintenance-free to the translator. Yes: Heavy maintenance is necessary with custom/proprietary MT engines, as their substantial feature is MT training, and here, programming skills are indispensable.
"Intelligence"/ Training	Static—the "Memory" feature: The database is enriched one step at a time. Where there are no similar or identical text chunks stored in memory, the technology will not help.	Dynamic: Publicly available MT engines work with dynamic text corpora reflecting recently available text material in the internet. The Training feature: MT engines can be trained (by use of algorithms) to avoid repeating errors. This can occur in real time.
Evaluation	No specific metrics to measure TM appropriateness, however, there are categories that help evaluate the usability of such tools (cf. above, section *The Comparison of CAT Tools*).	Yes: There are numerous metrics to measure MT quality, e.g. POST, BLEU, or TAUS Dynamic Quality Framework.
Harmonization	TM Exchange: An industry standard format was introduced to enable exchange of TMs between tools (the *.tmx format).	Neither the interoperability of MT engines has been harmonized nor the quality of their output. Quality standards for post-editing exist, e.g. KantanMT standards or TAUS guidelines.

The Process

Besides the systemic level, the two technologies differ in the aspect of involvement in the broader as well as narrower translation process.

TABLE 8.5 Comparing CAT and MT (PEMT) tools on the process level

	CAT	*MT / PEMT*
Where in the process	Suggested memory matches provide help in the translator's decision-making process.	Generated text enters post-editing as "raw material," influencing the final wording to a considerable extent.
On-Off feature	Yes. Translation memory can be turned off and translation can be performed from scratch.	Yes, in integrated interfaces of CAT tools. Dedicated post-editing interfaces, however, are not designed to bypass the MT functionality.
Eliminating redundancy?	Yes. Repetitive segments tend to be recalled from memory with identical wording as in previous occurrences.	No. Each text string is translated anew regardless of the fact if it has been translated before.
Assigning penalties	Yes, translation sources (translation memories as well as MT engines) can be weighted to select the most relevant/reliable one.	No. If multiple MT engines are used, preference is made only by defining the order in which the results appear (e.g. in a drop-down list).
Quality of original (Authoring)	Non-Key. If there is repetitive text, and hence a memory match, discrepancies of the original wording lower the efficacy of the system; yet, they can still be bridged by the applied fuzzy logic.	Key. Poorly spelled originals (e.g. OCR output) give rise to poor MT results. However, in some instances (e.g. foreign words, names), MT can overcome typos present in the original.
Productivity	Typically, there is a productivity increase when using TM technology. This is especially true with highly repetitive texts and dependent on the extent to which TM is populated with relevant text segments (TM saturation). However, when taken into consideration, labor-intensive pretreatment and posttreatment can reduce overall productivity gains.	Typically, there is a productivity increase when using MT technology. This is especially true with languages of large diffusion as well as non-creative, domain specific texts.
Concordance search	Limited to material available in translation memories.	Can include online searches in the MT environment.

The Product

In terms of the product, the following differences may be observed.

TABLE 8.6 Comparing CAT and MT (PEMT) tools on the product level

	CAT	*MT / PEMT*
Quality	Ranging from no-match instances (no text output from TM, hence quality considerations are irrelevant) to high or exact match scenarios (quality dependent on the quality of the retrieved text segments). Increases over time with TM saturation.	For example, depends on the volume of bilingual data available for MT creation and training. Ranging from prohibitive results with defective language to surprisingly good results (eventually "good enough") in specific language pairs, restricted domains, and/or from originals drafted in controlled language.
Context match	Yes; with individual segments, the system stores information on the preceding segment. In new occurrences, a check is performed to determine contextual relevance of a segment recalled from the translation memory.	No such feature under MT. Only domains can be determined, not the context.
Subject matter	Not relevant. It is repetitiveness that is critical.	Relevant, as—for subject-specific MT—it is specific text corpora that are of importance.
Text type	Very relevant, since some text types (e.g. administrative, technical texts) tend to be more repetitive than others are.	Not relevant, as the overall aim of rendering texts into a TL (regardless of its quality) can initially be arrived at with any type of text.
Terminology	Consistent: TM retrieves only what is in the memory. If a memory is created and maintained well, terminology consistency appears as a by-product.	Not consistent prior to MT training and PEMT. Neither SMT nor NMT technology distinguishes terms from general language.
Phrases	Consistent within high match-rate segments.	Not consistent, however idiomatic to a certain degree, due to the text-polishing phase of the intrinsic MT process, where the wording is matched against target-language text (TLT) corpora.

(*Continued*)

	CAT	MT / PEMT
Idiomatic expressions	Yes, as all text in a memory is produced by human translators.	Yes, to a limited extent. It depends if the engine was built with a preference for accuracy or fluency.
Can become counterproductive	Yes, where constituent preconditions are not met (e.g. non-repetitive texts) or pretreatment and posttreatment proves prohibitively challenging (e.g. alignment of low-quality OCR-generated texts).	Yes, with texts of a high degree of ambiguity (e.g. creative writing).

User's View

Technology alone is not the only relevant aspect. When comparing the technologies, one should also pay attention to the MMI and to user experience.

TABLE 8.7 Comparing CAT and MT (PEMT) tools on the level of user's view

	CAT	MT / PEMT
MMI	Ranging from a simple MMI to complex integrated environments.	Limited features: Dedicated PEMT interfaces typically lack advanced text-editing functionality.
User-friendliness	Limited: CAT tools are produced for specialized translators, not for general public. The technology is complex and the MMI is often not intuitive.	Limited: Whereas the MMI is not complex, usually some tutoring/training together with gradual experience is necessary to master such tools.
Psychology	Although the use of CAT tools has its impact on translation rates (high percentage matches and exact matches are priced less than text translated from scratch), there are few objections towards the technology, as it is just a tool to assist the professional translator, not to translate instead of him/her.	"Rage against the machine": MT can be perceived as a technology that steals work from translators.
Skill-degrading potential; risk of losing the translation competence	Low, as there will always be text segments that need translating (i.e. creating individual renderings), and not post-editing pre-processed TLT material.	Very high for translators, since the rendering phase is taken over by technology, and the translator turns to post editor of MT output.

	CAT	*MT / PEMT*
Ergonomics	Less harmful: As more writing and less editing is involved, more diverse operations are performed.	Potentially harmful: The manipulation with pre-translated text requires a limited range of repetitive input processes performed with high frequency. Occupational hazards may come as a result.
Source of Inspiration	Rarely: The process of automatic replacement of repetitive original text segments by translation text strings is a source of little surprise and/or creative solutions.	Yes, at times, as synonyms and varying plausible phrases can be generated based on the data that the MT engine draws on.
Wow factor	Limited, eventually negative, in cases where there is no memory match and the tool provides zero assistance to the translator.	Yes, triggered by unexpected equivalents or high-quality renderings.

9

THE FUTURE OF TRANSLATION AND TRANSLATORS IN A FAST-CHANGING ECONOMIC AND TECHNOLOGICAL LANDSCAPE

Giuseppe Palumbo

Introduction

The nature and scope of translation have been profoundly affected by both

- advances in information and communication technologies (ICTs) and
- processes related to globalization.

Today, virtually all instances of translation for purposes of publication (in print or online) involve computerized aids or digital resources of some kind. Currently in professional translation, "all processes are 'computer-aided' to some extent" (Pym 2014, 120). Even translation needs in everyday situations are increasingly met with the support of IT devices—typically an online electronic dictionary or a machine translation app available on a smartphone. In fact, in professional translation, computers and networks have swept away the idea of the translator as a solitary craftsman surrounded by piles of paper dictionaries. Likewise, ICTs have revolutionized the market for translation—often known as the translation or language-service "industry." What's more, all aspects of the field seem to be rapidly evolving. The translator's required skill set has changed markedly over the years. In response to technology advancements and other trends, new professional translation profiles have emerged, and the organization of labor has acquired traits typical of industrial sectors in terms of automation, scope of projects, and workflow (Shreve 2000; Dunne 2012).

At the same time, new business models, such as crowdsourcing, are poised to markedly influence professional translation practices in the future. Likewise, the platforms for translatable digital content are increasing in variety, and computers are getting better at assisting and, in some cases, replacing human translators. This is all happening at a time when the demand for translation seems to be growing—possibly even too fast with respect to the ability of the companies that provide language services to meet it (TAUS 2013). For technical communicators working in this context, knowledge is key to addressing this factor. Specifically, technical communicators need to develop a foundational grasp of the forces affecting modern translation practices to effectively engage in international content creation.

In this chapter, I address this need by examining the latest trends in the language service industry. In so doing, I illustrate the implications these trends have for individual translators and businesses (either offering or requiring translation and localization services). I also discuss prospective future developments in the industry. To achieve these objectives, I first discuss the impact of technology on translation. Next, I review recent developments in the language service industry. Finally, I conclude with thoughts on possible future developments that could affect both translation practice and the industry.

From Translation to Localization. And Back?

In translation as a commercial service, the widespread use of ICTs and digital content are factors that led to the emergence of an industry where the translation needs of companies (the creators of products or content) are met by language-service providers (LSPs)—other companies specializing in translation services. The linguistic and cultural adaptation of software and online content has become known as "localization."

In such situations, the use of "localization" is generally understood to refer to the processes and tasks leading to creating target-language version of a product or website. As an activity, "localization" encompasses both language transfer (applying concepts from one language to another) and engineering and management tasks (see Valli's chapter in this volume for more on localization).

Localization and the Project Approach to Translation Services

With the consolidation of the language service industry, many organizations have replaced the term "translation" with localization—the former being seen as an operation of language transfer taking place at a superficial level, i.e., not considering the rhetorical and cultural expectations of readers/users in the target language (to see how wrong such an understanding of translation is, see Minacori's chapter, this volume). The definitions of the term localization itself "tend to be contextually bound, reflecting the perspectives of those who formulate them" (Folaron 2006, 197). In fact, the term seems to be used with two distinct meanings (Pym 2011). More generally, localization is the overarching process of how a product or service is made to be linguistically and culturally appropriate to a target "locale" (i.e., a country or market defined by distinctive socio-cultural and linguistic features). Technically, localization is one step in the global product development cycle. In this cycle, localization is preceded by "internationalization," the process where the features of a product are rendered culturally neutral to facilitate subsequent adaptations (i.e., cultural custom fitting) to various locales. Items considered during internationalization include the separation of content from executable code and the adoption of standards for character encoding so that in target-language versions all characters display properly. Internationalization is, essentially, aimed at ensuring that technical adaptations are no longer necessary once localization starts.

The elements one needs to localize in an international product range from date, time, and number formats to color conventions and icons (Esselink 2000). The adaptation of such elements, however, is rarely the responsibility of translators, as localization (in the narrow technical sense) is usually taken care of by engineers and project managers. In particular, "localizable" elements relating to layout and graphics are usually addressed during the design stage of a product. The adaptation of other localizable elements, such as date, time, and number formats, may be left until the translation stage, when the dedicated text editors used by translators are able to automatically convert such elements into the appropriate target-language format.

For such processes—translation, the actual transfer of textual material—is typically outsourced to freelancers. In software localization, translatable text typically comes in resource files kept separate from source code. In website localization, translatable text often comes from authoring or content management systems. In both scenarios, translation memory tools are typically used to speed up the process. In particular, these tools reuse past translations either instantly within the current project or as leverage materials to be employed in future updates to a product or website.

An effect of the project approach to translation and the use of dedicated tools (e.g. translation memories) is that translators frequently encounter text fragments. Translation memories perform best on individual strings (e.g., phrases or sentences) of texts rather than longer passages (e.g., paragraphs). Because of this, project requirements often urge translators not to interfere with how a text is segmented by the tools in order to facilitate reuse for future projects. In addition, the focus on text segments generally means that translators work on individual strings removed from their original context. Often faced with decontextualized units, translators are bound to resort to phrase-to-phrase or even literalist equivalence strategies (Pym 2011, 6).

Web Localization

The separation of localization and translation is typical of large-scale localization projects, because their size and complexity require a clear division of tasks and roles and the adoption of well-organized protocols and procedures (Gouadec 2007, 309–310). In smaller projects, especially in web localization, the steps of product development may be different, leading to an overlap of functions and agents. Terminology management and translation, for example, may be assigned to different agents in larger projects, while in smaller projects they can be performed by the same people. Different degrees or levels of localization are also possible—especially in the case of websites (Jiménez-Crespo 2013, 33–36). In some websites, only a selection of paragraphs or pages may be translated into other languages. In other websites, several pages may have been localized, but navigation menus remain in the principal language.

Singh and Pereira (2005) have proposed a classification of localized websites based on the level of cultural adaptation, recognizing five degrees:

1. Standardized websites: sites in one language only
2. Semi-localized websites: sites only providing contact information about foreign subsidiaries to the international audience
3. Localized websites: sites for which most content has been localized without affecting the site structure
4. Highly localized websites: sites that have been localized and fully adapted even in their deep structure
5. Culturally adapted websites: sites that are totally immersed in the target culture

Also, as Jiménez-Crespo (2013, 35) notes, the higher the degree of localization, the more likely that the deep structure of the website must be re-engineered.

Extensive re-engineering may require the adoption of more complex and expensive integrated systems that manage authoring and translation workflows. Therefore, cost considerations affect a company's decision to adopt one or the other level of localization. A classification like the one proposed by Singh and Pereira (2005) seems to be particularly suitable for analyzing the localization strategies of multinational companies. These companies are likely to have both a commercial interest in reaching international customers and the ability to devote enough financial resources to doing so. For multinationals, in other words, the investment of time, money, and resources in developing a truly culturally localized site may pay off spectacularly.[1]

Other recent models of localization (such as those based on the involvement of end users), however, show cultural adaptation is not necessarily the focus of the overall localization strategy. Rather, the focus can be on the immediate availability of translated content or the need to ensure usability. In this respect, a broad distinction between two categories of websites is possible:

1. Websites offering services envisioned, designed, and developed expressly for the web (i.e., "web-native" platforms or services): This category includes commercial services, such as Skype, and platforms or websites offering free services but relying on advertising for revenue, such as social networking or community sites (e.g., Facebook and Twitter) and search engines (e.g., Google and Bing).
2. Websites stemming from traditional business or communication models and adapted to online environments to varying degrees: This second category would include a variety of models and types from online stores to the corporate websites of large multinationals.

In terms of localization, various web-native services can be seen to have adopted models that do not completely fit classifications based on the degree of cultural adaptation such as Singh and Pereira's. Rather, for such web services, translation seems to have acquired a different significance. Its primary purpose is enabling users to instantaneously *use* services, as opposed to extracting information from them.

Web-native services do not seem to be the expression of any given national culture. Rather, they can be seen as the product of a shared generational culture that is accustomed to communicating and socializing online. In this context, translation becomes a means or "utility" (TAUS 2013) to be employed to overcome the last true barrier to accessibility: language.

In many cases, web services have become much more "user-centric" than "producer-centric." This shift is having an impact on the models that service providers use to translate content (Garcia 2009). In particular, the growth of web-driven demand for translation is exposing the limitations of traditional localization models. Such models are often considered too costly and time-consuming, especially for situations where instant accessibility of content is of paramount importance. If alternative modes of translation generation are seen to lead to acceptable levels of quality, then they are preferred over more traditional localization models.

The Language Service Industry: A Brief History and Recent Developments

Outsourcing is still the dominant model for addressing translation and localization in today's language service industry (Dunne 2012). Organizations needing localization services outsource these activities to an LSP. In turn, the LSP may subcontract translations and other related services to another LSP specializing in a specific language and working with in-house staff and freelancers, or at times with freelancers only. Work is organized in projects, and specific software tools are available to assist members of a team in their specific activities and tasks. Alternative models of translation generation are, however, emerging to take advantage of advances in machine translation (MT) or the possibility to actively involve end users.

The Emergence of Dedicated Translation Tools

Software localization has been the first sector to heavily use computer tools for translation and other related activities. Two essential factors have contributed to this:

1. the digital nature of translatable content
2. the particular nature of the texts to be translated

A software "product" has traditionally been understood to comprise the software itself, the product documentation, and promotional materials (e.g., packaging and advertisements). The translatable software content (i.e., the user interface, graphics elements, scripts, and any other visible text) is digital by definition. In the early days, the other materials were mostly available in print. Over time, they have increasingly been made available in digital format. Today, most software products are associated exclusively with documentation in digital format. This documentation is typically available as documents, online help,[2] or dedicated websites organized as tutorials or knowledge bases.

In localization, the text to be translated has traditionally been available largely in digital format; today, however, it tends to be exclusively digital. Software applications constituted the bulk of localization work in the early days of the industry. Now, a wider array of products and services embed significant amounts of digital content, making them amenable to localization. The list includes websites, videogames, smartphone apps, small devices (such as mobile phones and MP3 players), and multimedia platforms (Jiménez Crespo 2013, 28–29). Additionally, the incessant development of technology leads to a continuous redefinition and blurring of the lines of demarcation between such products or devices.[3] Increasingly, gadgets or devices are turning into "digital touch points" that enable users to access personalized content. This factor means the demand for translation is likely to grow exponentially as technologies provide users with new and different ways to access and use texts (TAUS 2013, 10).

Historically, the availability of translatable content in digital format was not the only characteristic that favored the introduction of dedicated computer tools for translators (i.e., tools offering special capabilities beyond word processing). Repetitiveness of both vocabulary and sentence structure was the other key aspect. Most text associated with a software product tends to come in short chunks (typically the text strings in user interface elements, such as commands and menus, error messages, and dialog boxes) or in longer stretches (the sections and paragraphs of the product documentation). As a high degree of repetition characterizes the shorter chunks and the longer stretches of text within a product, organizations soon realized the productivity of translators could increase by developing a tool capable of

- recognizing previously translated segments of text housed in a central, searchable database and
- offering translators the possibility of reusing past translations in that database—or obtaining suggestions from them based on similar texts found in that database.

The tool that put this principle into practice was translation memory (TM). Made commercially available in the 1980s, TM has been increasingly adopted by the language-service industry since its inception. A TM is, strictly speaking, an electronic database containing SL (source language) and TL (target language) segments to be reused for new translations.[4] TM tools were originally used only in localization projects. They were later embraced in other areas of professional translation by both LSPs and freelancers. Their use was particularly prevalent in situations where

- text was available in electronic format,
- content was characterized by the repetitiveness of language structures, and
- frequent content updates made reuse of past translations worthwhile.

An interesting feature available in many TM tools was their capability to support text in a wide variety of file formats. This feature allowed freelancers and LSPs to dispense with buying the corresponding software applications because the specific editing and layout features of a given format were taken care of by the TM tool itself. For LSPs, a crucial development was the possibility to store TMs on a local server accessible to multiple users. The crux of this benefit was that LSPs could retain all of the strings (i.e., text) translated using a TM tool, thus gaining complete control over the previously translated material to be made available at the start of a new translation project.

The use of TM tools brought about several benefits for the industry. Productivity increased significantly as large amounts of already translated text ("legacy" materials or "language assets") could be reused from previous, related projects. The possibility of storing legacy materials on central servers gave LSPs total control over them. Additionally, access to the same TMs by the team of translators working within the same project led to increased consistency in the translations they proposed. The integration of TMs with termbases (i.e., databases specifically meant to store terminological entries) also increased consistency. Finally, TM tools were gradually designed to incorporate functions of quality assurance and project management. This development made it possible to obtain statistics on new translatable content in relation to legacy materials. (Such statistics are particularly important during project preparation, e.g., for purposes of scheduling, costing, and quoting.) All of these factors made TM the most significant tool in the language-service industry in the three decades starting from the 1980s.

The Integration of Translation Memory and Machine Translation

Whereas TMs were initially located on individual machines, they were later moved to an LSP's central corporate server. This move made them simultaneously accessible to multiple users and managed (and owned) by the LSP. Gradually, TMs and glossaries have come to be integrated in overall translation management systems used to perform the tasks and functions pertaining to a localization project. Within these systems, one can establish specific procedures for leveraging past translations, integrating available tools, and the assuring quality. In particular, the integration of tools may lead to workflows organized as follows (and as depicted in Figure 9.1):

- A new translatable text is first pre-translated on the basis of TM.
- The pre-translated text is then processed with MT so as to fill in the segments left untranslated by the TM.
- The translated text is revised, or "post-edited," by a human translator.

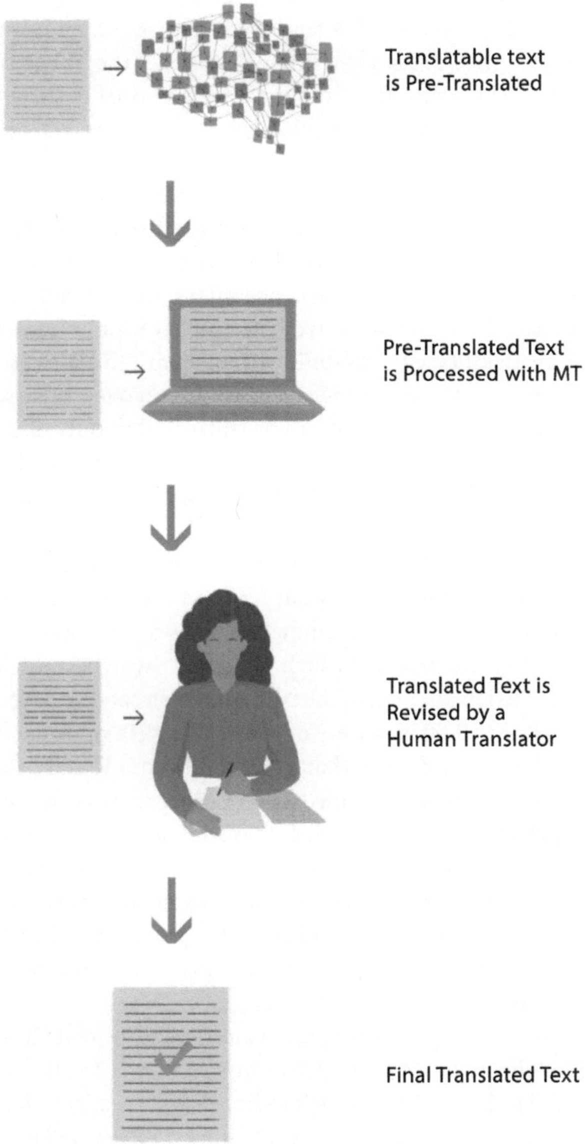

FIGURE 9.1 Integration of TM and MT in the translation process

This workflow follows an approach sometimes referred to as "enhanced TM" (Garcia 2012), currently researched and implemented by several developers of TM solutions. It has also been applied to online platforms, such as Google's Translator Toolkit.

With the integration of TM and MT, human intervention focuses predominantly on post-editing the results of such processes. In these cases, the translator reviews and edits the text resulting from translations done by a machine. Recent developments take this integration even further. Newer versions of SDL Trados Studio 2019® (a popular TM management system), for example, include a new technology, called AdaptiveMT, which changes the MT output based on the post-edits made by users. Thus, the software learns from the user's post-edits of the automatically translated segments and adapts the MT output for similar source segments encountered later on in the text.

Another approach to improving the performance of translation tools is based on the "pre-editing" of texts based on the rules of controlled language. Controlled languages require individuals to write in a certain restricted way involving the use of a specific (preset) vocabulary and restricted grammatical and stylistic constraints. The objective is to improve the translatability, comprehensibility, readability, and usability of an item. The idea is to limit the meaning of a word, a phrase, a sentence, and even an overall passage to a very specific definition—ideally, one that can be translated in only one way. Various controlled language rule sets exist for English, and these were developed by a variety of entities.[5] (An example is the ASD Simplified Technical English, developed by the aerospace industry.)

The sets of rules for a controlled language specify generally which grammatical structures are allowed and what terms can be used to achieve the intended purpose. For example, the guidelines for several controlled languages recommend that relative pronouns such as "who," "which," and "that" should not be omitted (O'Brien 2003). Another different form of controlled language is Global English (Kohl 2008), a variety of written English optimized for a global audience according to guidelines that go beyond those found in conventional style guides. The guidelines are aimed at rendering texts easier to translate or easier for non-native speakers of English to read. For example, an author might write, "You can also conserve storage space by compressing files," instead of, "Another way of conserving storage space is compressing files." Such phrasing clarifies that "compressing" is a gerund and not a present participle (Kohl 2008, 149). Global English also emphasizes grammatical structures and terms to be avoided rather than those that are allowed.

The application of controlled language rules on source texts has been shown to have positive effects on MT in terms of quality (see, for instance, Nyberg, Mitamura, and Huijsen 2003), acceptability by TL readers (Roturier 2006, quoted in O'Brien 2012), and post-editing effort (O'Brien 2006). More recent research (e.g., O'Brien 2010), however, indicates that using controlled language rules in texts to be machine-translated does not necessarily lead to target texts associated with higher readability scores.[6] Within this context, Thicke (2011) describes the effect of using rules of Global English for pre-editing and terminology adaptations in MT. She concludes that "untrained MT is two times faster to post-edit than to translate from scratch; trained MT is three times faster; and trained MT with source control is four times faster" (Thicke 2011, 40).

Another recent development of translation management systems is their migration to cloud-based services. Some established TM developers and a few newer entities have started offering cloud-based platforms providing the software infrastructure needed for the completion of a localization project. The service is aimed at either LSPs or translation buyers that decide to exploit their language assets independently or project manager who assigns translation work to freelancers allowed to log in to cloud-based TMs and resources.

Consequences of the Large-Scale Adoption of Translation Memories

For individual translators, the consequences of the large-scale TM adoptions were rather controversial and have been discussed by researchers, translation scholars, and industry commentators (e.g., Biau Gil and Pym 2006; Garcia 2009, 2012). From the perspective of translation as a language transfer operation, the foundational principle on which TM tools are based makes the translator focus on restricted units of language. This situation is compounded by the use of content management systems (CMSs) in which content is also frequently segmented to facilitate changes and updates.

In short, the introduction in the translation workflow of TMs and CMSs often leads to decontextualization of the units needing translation. The effect, sometimes referred to as "chunking," is a product of how digital content is created and managed even before the translation stage. Such factors cause translators to focus on shorter chunks of language and ultimately abandon linear readings of what they translate. Accordingly, this model appears in stark contrast to the top-down approaches to translation favored in scholarly research and in many translator training programs.[7]

New Notions of Text and Content in Digital Environments

The renewed emphasis on equivalence at a restricted level may be linked to a redesign of the notions of text and content in digital environments. In printed media, both the production and reception of texts had a certain degree of linearity. With digital content, text creation and reception do not usually proceed in linear fashion. After the release of an initial version, digital content goes on being generated in successive rounds of updates and modifications. Content, in other words, becomes "dynamic," which implies working on a continuous stream of small files or even individual strings vs. organizing large-scale projects. In such cases, the immediate availability of the resulting translation becomes the most important objective (Garcia 2012).

In many platforms, users are responsible for content creation, and user-generated content lies at the heart of many online services and platforms. Where such content is mostly textual (as in TripAdvisor reviews), it then needs to be translated. Similarly, the reception of digital materials, especially in online environments, rarely presupposes readings that proceed from a beginning to an end. In fact, when seeking procedural information online, individuals are likely to do dialogic searches where they interact with machine retrieval systems (as in knowledge bases) and other users (e.g., online forums).

Alternative Models of Translation Generation

The interactional model of knowledge extraction and exchange has entered the realm of translation *generation*, giving rise to models that provide alternatives to traditional localization processes. One such model is crowdsourcing, or the use of volunteer translators (DePalma and Kelly 2011). Web-native services, such as Facebook and Skype, have adopted this model and enlisted their own users as translators and organized them into communities with varying degrees of exclusivity or openness (described in detail in Mesipuu 2012), with the possible supervision of in-house language specialists checking terminology or performing quality control. The principle behind crowdsourcing is that the target language version is created through negotiations among the community of users. It is a process in which the users' collective translation effort and their in-depth knowledge of the service are considered enough to reach an appropriate level of quality. In this way, translation work might focus on product features and elements deemed of major importance. In such situations, lower-risk elements are often either translated inaccurately or not even translated at all (see Figure 9.2).

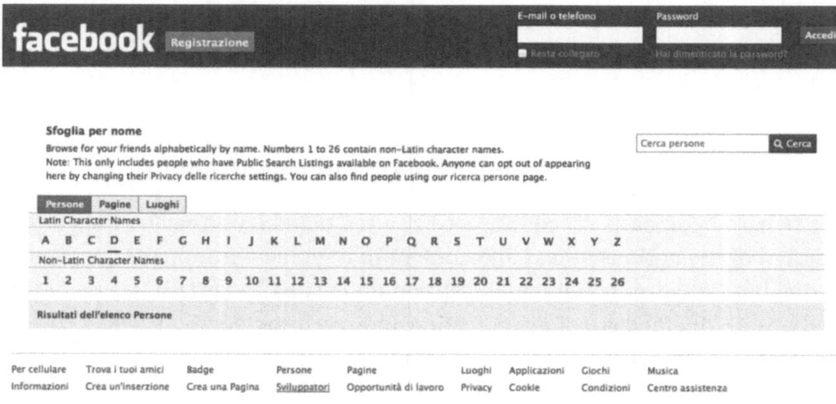

FIGURE 9.2 Page from the Italian Facebook website (visited on 15 September 2013) showing a paragraph with text shifting between English and Italian

It is not difficult to imagine that the non-translated strings found in the Facebook page shown in Figure 9.2 would be considered a serious error. However, the gravity of the error acquires a new dimension within the model of translation generation chosen by the site. Other, presumably higher-risk elements of the interface have received more attention by the community of volunteer translators. Additionally, the site managers appear to accept lapses or inaccuracies that do not affect the overall user experience. Their view of quality seems to accommodate factors different from those contemplated in established quality evaluation models. As Garcia (2012) reminds us, in dynamic content situations, "mistakes will need fixing only after the user community has discovered them" (457), which presumably happened with the Facebook page in question.

Translation Quality: An Evolving Notion

Quality has received much attention from industry and scholarly research, but the issue has remained controversial, due to a lack of common ground among practitioners, commentators, and researchers. A survey conducted within the OPTIMALE project,[8] showed the list of top-ranking competencies and skills requested of translators by LSPs was led by the ability to produce "100% quality" (OPTIMALE 2011, 7). However, it is difficult to say how quality would have been defined by LSPs if they had been asked to spell out the respective role of cost, time, and text accuracy in their definitions. Had translator trainers taken part in survey, the result would have been the same, but the implied notion of quality might have been sensibly different.

Views on quality in translation vary depending on whether translation is seen as a product, a process, or a service. The approaches to quality proposed in translation studies or adopted in the translation field are based on one or the other view. Scholarly research has traditionally focused on translation as product (i.e., on translated texts). Quality was assessed in relation to the degree of equivalence between target translation (TT) and source text. Assessment criteria were then extended to account for the variety of factors that constrain translators' choices in the process of translation. Such factors range from the function of texts (House 1977; 1997) through to the expectations of TT readers and the norms that govern the activity of translators (Chesterman 1997).

In industry, translation is seen mainly as a service, which leads to an emphasis on procedural aspects when quality is assessed. Adopting translation procedures that accommodate how tools process texts is thus of paramount importance as translators employ these tools to increase productivity. Quality standards for translation are mostly related to the process of translation: They establish procedural guidelines and attempt to ensure customer satisfaction. Existing standards include the J2450, developed by the Society for Automotive Engineers (SAE), and the ISO 17100:2015 (formerly EN-15038), a European standard that defines the requirements for the provision of quality services by translation vendors.

In general, the standards do not establish criteria for measuring quality in the translated texts. Rather, they focus on the correctness and reliability of the procedures adopted to arrive at texts. The assumption is that if a system of accountable and transparent quality control for the translation process exists, the product will be qualitatively acceptable (i.e., capable of satisfying the agreed customer requirements). For instance, the EN-15038 establishes that any translation service must include a review stage at a minimum and that translators should possess certain predefined qualifications. Thus, the emphasis of quality evaluation in industry is on the ability of processes to reach a compromise between acceptable quality for end users and constraints of cost, time, and resources.

Quality Assurance (QA) is the label usually given to the procedures applied throughout the translation process to ensure the quality requirements of clients are met. These procedures focus on ensuring quality of service (e.g., in terms of deadlines), quality of presentation (e.g., in terms of layout and format), and linguistic quality.

Quality Control (QC) is the common term for the procedures adopted to identify problems in randomly selected passages of a translated text. These procedures help evaluators determine the degree to which the text meets professional standards, the standards of the translation organization, or customer requirements. The most popular model for quality control procedures remains the model developed by the now-defunct Localisation Industry Standards Association (LISA). In the LISA model, the evaluation of translated texts is carried out through an error analysis based on a typology that covers language, formatting, and functionality errors and assigns different levels of severity to each error.

Translation quality evaluation based on models such as LISA's has been criticized as being mostly based on a static and serial model of translation production (O'Brien 2012). More specifically, O'Brien sees existing quality evaluation models as unable to respond to a series of new concerns expressed by translation buyers. These concerns have to do with aspects such as the

- need to consider customer opinion and feedback,
- relevance of new forms of web-mediated communication,
- perception that existing quality evaluation models are expensive and time-consuming, and
- need to make room for newer forms of translation generation such as the integration of TM and MT.

For O'Brien, more dynamic models of quality evaluation should assess translations according to

- utility (i.e., how effective the translated text is from a functional point of view),
- time (i.e., the ability to produce translations at the required speed), and
- sentiment (i.e., the impact translated text has on brand image).

Individuals can consider each aspect "most important" in evaluation, depending on the type of content and the type of "communication channel" (i.e., who the primary addressees of texts are). User interface text and user documentation, for instance, would rate higher for utility, while marketing materials and website content would rate higher for sentiment. O'Brien also proposes that quality evaluation consider aspects such as control over the content creation process in the SL according to Global English guidelines like those given in Kohl (2008).

In summary, the diversification of content platforms and translation generation modes is leading researchers and practitioners to research new models of quality evaluation with the objective of providing "adequately targeted quality" (TAUS 2013, 61). In these new models, some factors are likely to receive increased attention. They include quality expectations on the part of end users, the degree of risk or criticality associated with particular content types, and, crucially, the quality of source texts, which should facilitate the work of both human translators and computer tools.

Future Trends in the Practice of Translation

These technological and digital content advances mean current translation practices are undergoing a profoundly transformative stage. In the last three decades, translation technology was essentially developed in the interest of the localization industry and with the primary aim of improving the productivity and consistency of human translation. Over the last few years, there has been an explosion of digital content. Additionally, new players have entered the arena of translation technology, starting a process that is rearranging the translation market.

The future of this process is not easy to predict, due to its connections to developments in information and communication technology. (This area is notoriously difficult to study in terms of making projections, especially for the long term.) However, some trends may indicate how the translation industry and the practice of translation might evolve in the coming decade. Translations will be produced with major changes in two key respects:

1. who translates, how, and why and
2. how machines are integrated in the process of translation.

Thus, we can hypothesize the following future developments.

Who Translates, How, and Why?

Translation as a professional activity has really emerged only in the last 50 years. Professionalization here is based on the perception that the activity requires certain technical skills, that it is increasingly institutionalized, that it adopts quality control systems, and that it relies on accreditation procedures (although, in the particular case of translation, these are rarely binding). A professional translator can be defined as someone who regularly carries out remunerated translation work, or someone "who is a translator," as opposed to "someone who does translation (sometimes)" (Chesterman 2001, 146).

The widespread dissemination of computers and networks and the explosion of digital content are changing all this. Demand for translation is growing so rapidly that professionals seem unable to satisfy it. Tool-based human-computer interaction and resources are putting many people with sufficient bilingual competence or keen subject-matter specialty in a position to generate translations. Users of web content are also becoming producers of content, and production in many cases includes translation. Fortunately, machine translation is improving, thanks to the increasing availability of "translation data." (Such data are parallel text corpora accumulated from translation memories and online translation service platforms or harvested from localized versions of websites and other sources [TAUS 2013, 69].)

New translation generation models have emerged alongside institutionalized localization models developed by the language industry over the last three decades. These models generally rely on online collaboration and take advantage of the willingness of people to "work as translators (sometimes)," mostly on a voluntary basis. The general terms used for such models are "crowdsourcing" and "community" or "collaborative" translation (see O'Hagan 2011). All refer to situations where "a company or institution purposefully puts out a call to the community to complete a specific localization task" (Jiménez Crespo 2013, 193). This definition also includes translations solicited by nonprofit organizations. The broader term "user-generated translation" (O'Hagan 2009) encompasses such situations plus cases where translation is not solicited but performed voluntarily by self-selected groups who wish to distribute translated materials.

The common denominator for these collaborative translation forms is that they make ample use of the possibilities afforded by digital communication. They do so either in terms of the possibility to negotiate translation solutions or for the dissemination of translated content. As to the purposes of crowdsourced translation projects, Ray and Kelly (2011) have identified three main categories:

- "cause-driven" projects, that is, nonprofit initiatives motivated by a common goal (e.g., of a humanitarian nature);
- "product-driven" projects (i.e., those aimed at filling gaps in the localization of specific products, for instance by translating for open-source software or providing a version of a commercial product in a language not made available by the producers); and
- "outsourced-driven" projects, that is, those started by for-profit companies.

Volunteer translators in projects belonging to this last category have been observed to be motivated by "productive selfishness" or the desire to improve a product or service with which they are very involved, or, more generally, a desire to realize one's potential (Mesipuu 2012).

What are the implications of these new models for practicing professional translators? The traditional translate-edit-proofread (TEP) model (Beninatto and DePalma 2007, quoted in Garcia 2012; see also Minacori's chapter, this volume) will go on being used for products with longer shelf life or in cases where brand image, legal requirements, or prestige suggest minimizing the risk of getting translation wrong. In other cases, professional translators may have to accept a redefinition of their job profile. The increasing integration of MT into the translation process is already giving prominence to post-editing (i.e., the editing performed on a text translated by an MT system), which for some translators may become the bulk of their activity. Possible new key figures will be the technical writers working on controlled source text to feed into MT and data base managers and reviewers tending translation data (Garcia 2009). New specializations linked to the emergence of the previously mentioned collaborative models may consolidate. For example, the roles of community translation manager and crowdsourcing quality manager may become one (Jiménez-Crespo 2013: 192).

Another possible development for professionals is a renewed impetus on translation into the second language (i.e., the translator's non-mother tongue). A taboo according to professional codes of conduct in some countries, translation into the translator's foreign language is a reality in many smaller markets where demand does not meet supply (Pokorn 2005), for example from Maltese to Finnish. The increased availability of MT and translation data—and the renewed expectations on quality—may favor this trend especially in certain language pairs (and particularly with English as a target language).

How Will Machines Be Integrated in the Process of Translation?

Given all of the previously mentioned developments, the question should perhaps more appropriately read: "How will machines be *further* integrated in the process of translation?" The general tendency is for a progressive shift towards human-assisted machine translation, with continued reliance on TM. The two most popular architectures for MT systems (rule-based MT and statistical MT) have both proven suitable for such integration, particularly where MT engines can be customized. Another, more recent architecture, called "neural machine translation," has been shown to perform even better than the other MT architectures (Wu et al. 2016) and is now the basis for Google Translate, which previously relied on a phrase-based statistical algorithm.

Free MT is still not likely to be considered suitable for texts destined for dissemination, especially where brand image is at stake. However, where the objective is assimilation—in other words, translation generated by users for their own consumption—free MT is already today considered a viable option. In fact, it is the first port of call for web users who want the gist of a foreign web page they are navigating.

In industry, the process will most likely be where technical communicators pre-edit texts and translators post-edit MT output. For less critical content, either pre-editing or post-editing will probably suffice. Availability of high-quality translations produced, or post-edited, by humans will still be a crucial element, given that MT systems, especially statistical systems, are improved on the basis of the translation data (essentially, parallel corpora) fed into them.

Dedicated translation tools such as TM and terminology management systems will continue to be refined. Additionally, the data that they create and the management components themselves will increasingly be stored in cloud environments. The client/server model is already being replaced with a cloud-based service model. Hosted services present several advantages over software installed on customer premises (TAUS 2013, 51):

- a continuous software release cycle that eliminates the need for upgrades
- reduced system administration and IT costs
- the possibility to adopt subscription-based cost models, as opposed to the more expensive perpetual licenses
- accessibility across a broader range of devices and operating systems
- improved user interfaces and usability

Moreover, as the availability of translation data increases and translation data storage moves to centralized environments, a key aspect will be the improved effectiveness of "leveraging" (i.e., the reuse of past data for translating new content). In this respect, improvements in matching at the sub-segmental (or phrase) level of translation units will be sought to further increase the effectiveness of TM leveraging and the productivity gains it leads to. In effect, a point may someday be reached where the boundary between TM and MT (already a fuzzy one) disappears completely.

Human intervention will still be crucial to provide precise quality measures or to supervise automated process. In particular, professional translators are likely to take on other roles besides that of MT output post-editors. These roles will include the

- assessment of both MT output and the corpora used to train MT engines,
- editing of these corpora,
- management of terminology,
- pre-editing of source texts, and
- refinement of user interfaces according to translator needs (O'Brien 2012, 119–120).

Finally, the general trend that probably encapsulates all those described so far is the increased availability of language data—particularly translation data—in online environments. This item is the single aspect that may prove to be the most influential development in the way translation services are provided over the next decade or so. The process becomes a virtuous circle if the data fed into MT engines come from quality translations that are then turned into well-tended parallel corpora. It may, however, also turn into a vicious circle if the input fed into MT systems (or even TMs) is not of the desired quality, thus leading to a propagation of poor translations. Another problem with the collection of translation data is that harvesting them freely from the web may lead to copyright infringements, which means that dedicated repositories might be a better choice for developers of MT who want to train their systems.

A Concluding Note

A few years ago, MT and translation-oriented computerized technology in general were already described as a complex and diverse field that sees "the involvement of a wide range of 'agents' from researchers and tool developers, through evaluators to various end-user groups including professional translators, trainers and translation companies" (Quah 2006, 196). In the intervening years, opportunities for interaction between these agents have been created (e.g., associations uniting practitioners, researchers, and trainers, such as the Translation Automation User Society or TAUS). However, research on language technology and MT and research on translation studies have proceeded on fundamentally separate tracks.

Considering the Future

Technologists have forgotten to involve practitioners or scholars who might have provided informed feedback on their research. Similarly, no proper "technological turn" (O'Hagan 2012) has been observed in translation studies. While the pace of technological advances in translation is relentless, there is still a need for experts in language and cultural transfer. How such expertise will be put to use in the future may be markedly different from the ways in which it has been put to use in the last few decades. Research combining the technological and the translation studies camps may help better redefine expertise in language and cultural transfer.

This chapter proposed a selective, and probably biased, overview of the trends reshaping the today's translation industry and translation as a profession. These trends essentially result from the ubiquity of digital content over a wide variety of platforms and devices. They also result from the pressing need to overcome language barriers in accessing such content. An attempt has also been made at anticipating future trends, in the awareness that the fast pace of technological change may render any such attempt useless, if not ridiculous.

Whatever the future may hold, one certainly positive effect of the increasing use of digital technology in translation, to which language users have become much more alert, is its relevance and significance. Especially in largely monolingual societies, until very recently language transfer skills and activities were either taken for granted or considered to be the remit of a few fortunate bilinguals (which, of course, ignores the fact that humanity has always included a very large segment of individuals who could fluently speak more than one language). The rise of digital content and the need to access it instantaneously have made episodes of communication breakdown due to poor translation more conspicuous. This situation has led language users at all levels to realize what a delicate and, at the same time, powerful tool language is in itself.

A Summary for Professional Communicators

As alternatives to localization, new models are emerging for the generation of translated content. They will likely consolidate in the near future. These models are based on an increased use of machine translation (MT) and the involvement of end users (e.g., in translation "crowdsourcing"). As content itself becomes more dynamic and "user centric," instant access to translated versions will be considered a key aspect. This situation will affect how the quality of translated content is evaluated. Translators, in turn, will need to consider different parameters according to the type of content to be translated. In technical documentation, the usability of translated content is likely to play an increased role, thus complementing the current focus on readability and procedural correctness.

Translation needs will increasingly have to be taken into account at the stages of content design and document drafting (e.g., through refinements to user interfaces and the adoption of Simplified English rules). Doing so is necessary to facilitate the work of either human translators or MT systems carrying out the actual translation task. Increasingly, source language quality will be recognized as having a major impact on translation quality, and technical communicators will be involved in the pre-editing of content to be fed into MT systems.

The increased availability of language data, and particularly translation data, in online environments is likely to be the most influential development in the way translation services are provided over the next few decades. MT engines are trained on available translation data. The resulting increase in the volume of translations available in digital format means that machines have at their disposal more data for training themselves. Both professional translators and professional communicators are likely to be involved in projects aimed at tending translation data so that MT output remains of consistently good quality.

Notes

1 Interestingly, Singh and Pereira found no "culturally customized Websites" in their 2005 survey of 307 US multinationals, but a replication of the survey may give different results today.
2 Online help, which was originally stored entirely on an individual machine, is usually available today in the software producer's website.
3 For example, almost all mobile phones have today morphed into smartphones, and most of these already act as multimedia platforms.
4 The term "TM" is sometimes used to refer to the program that exploits or manages the database—or what we might call a "translation memory system" or "manager."
5 An overview of the notion of controlled language, with several examples, is given in O'Brien (this volume); O'Brien discusses how professional communicators can write keeping in mind the needs of translators and machine translation systems.
6 For studying the comprehensibility of such texts, usability testing would probably be more appropriate than readability scores.
7 In the last few decades, translation studies scholars have strived to construct models of quality taking into account equivalence at text level and, in some cases, giving translators the role of text (re)designers in the TL. By contrast, the use of translation tools in industry has reaffirmed the primacy of equivalence at sentence or even at word level.
8 OPTIMALE was an EU-funded project (2010–2013) aimed at enhancing the visibility and relevance of professional translator education and training in Europe. Further information on the project can be found at www.translator-training.eu

References

Beninatto, Renato S., and Donald A. DePalma. 2007. *Collaborative Translation: The End of Localization Taylorism and the Beginning of Postmodern Translation*. Lowell, MA: Common Sense Advisory.

Biau Gil, José Ramón, and Anthony Pym. 2006. "Technology and Translation (A Pedagogical Overview)." In *Translation Technology and Its Teaching*, edited by Anthony Pym, Alexander Perekrestenko, and Bram Starink, 5–19. Tarragona: Intercultural Studies Group.

Chesterman, Andrew. 1997. *Memes of Translation: The Spread of Ideas in Translation Theory*. Amsterdam: John Benjamins.

———. 2001. "Proposal for a Hyeronimic Oath." *The Translator* 7, no. 2: 139–154.

DePalma, Donald A., and Nataly Kelly. 2011. "Project Management for Crowdsourced Translation: How User-Translated Content Projects Work in Real-Life." In *Translation and Localization Project Management: The Art of the Possible*, edited by Keiran. J. Dunne and Elena. S. Dunne, 379–408. Amsterdam: John Benjamins.

Dunne, Keiran. J. 2012. The Industrialization of Translation: Causes, Consequences and Challenges." *Translation Spaces* 1, no. 1: 143–168.

Esselink, Bert. 2000. *A Practical Guide to Localization*. Amsterdam: John Benjamins.

Folaron, Deborah. 2006. "A Discipline Coming of Age in the Digital Age: Perspectives on Localization." In *Perspectives on Localization*, edited by Keiran. J. Dunne, 195–222. Amsterdam: John Benjamins.

Garcia, Ignacio. 2009. "Beyond Translation Memory: Computers and the Professional Translator." *JoSTrans: The Journal of Specialised Translation* 12. www.jostrans.org/issue12/art_garcia.php

Garcia, Ignacio. 2012. "Machines, Translations and Memories: Language Transfer in the Web Browser." *Perspectives: Studies in Translatology* 20, no. 4: 451–461.
Gouadec, Daniel. 2007. *Translation as a Profession*. Amsterdam: John Benjamins.
House, Juliane. 1977. A Model for Translation Quality Assessment. Tübingen: Narr.
———. 1997. Translation Quality Assessment: A Model Revisited. Tübingen: Narr.
Jiménez-Crespo, Miguel A. 2013. *Translation and Web Localization*. Abingdon: Routledge.
Kohl, John R. 2008. The Global English Style Guide: Writing Clear, Translatable Documentation for a Global Market. Cary, NC: SAS Institute.
Mesipuu, Marit. 2012. "Translation Crowdsourcing and User-Translator Motivation at Facebook and Skype." *Translation Spaces* 1, no. 1: 33–53.
Nyberg, Eric, Teruko Mitamura, and Willem-Olaf Huijsen. 2003. "Controlled Language for Authoring and Translation." In *Computers and Translation: A Translator's Guide*, edited by Harold Somers, 245–282. Amsterdam: John Benjamins.
O'Brien, Sharon. 2003. "Controlling Controlled English: An Analysis of Several Controlled Language Rule Sets." Paper presented at the Joint Conference combining the 8th International Workshop of the European Association for Machine Translation and the 4th International Workshop on Controlled Language Applications, Dublin City University, Ireland, May 15–17.
———. 2006. "Machine-Translatability and Post-Editing Effort: An Empirical Study Using Translog and Choice Network Analysis." Ph.D. dissertation, Dublin City University.
———. 2010. "Controlled Language and Readability." In *Translation and Cognition: ATA Scholarly Monograph XV Series*, edited by Gregory Shreve and Erik Angelone, 143–165. Amsterdam: John Benjamins.
———. 2012. "Towards a Dynamic Quality Evaluation Model for Translation." *JoSTrans: The Journal of Specialised Translation* 17. www.jostrans.org/issue17/art_obrien.pdf
O'Hagan, Minako. 2009. "Evolution of User-Generated Translation: Fansubs, Translation Hacking and Crowdsourcing." *Journal of Internationalization and Localization* 1: 94–121.
———. ed. 2011. "Community Translation 2.0." Linguistica Antverpiensia, New Series: Themes in Translation Studies 10.
———. 2012. "The Impact of New Technologies on Translation Studies: A Technological Turn?" In *The Routledge Handbook of Translation Studies*, edited by Carmen Millán and Francesca Bartrina, 503–518. Abingdon: Routledge.
OPTIMALE. 2011. *The OPTIMALE Employer Survey and Consultation*. www.ressources.univ-rennes2.fr/service-relations-internationales/optimale/attachments/article/52/WP4_Synthesis_report.pdf
Pokorn, Nike K. 2005. Challenging the Traditional Axioms: Translation into a Non-Mother Tongue. Amsterdam: John Benjamins.
Pym, Anthony. 2011. "What Technology Does to Translating." *Translation & Interpreting* 3, no. 1: 1–9.
———. 2014. *Exploring Translation Theories*. 2nd ed. Abingdon, UK: Routledge.
Quah, Chiew Kin. 2006. *Translation and Technology*. Houndmills, UK: Palgrave Macmillan.
Ray, Rebecca, and Nataly Kelly. 2011. *Crowdsourced Translation: Best Practices for Implementation*. Lowell, MA: Common Sense Advisory.

Roturier, Johann. 2006. "An Investigation into the Impact of Controlled English Rules on the Comprehensibility, Usefulness and Acceptability of Machine-Translated Technical Documentation for French and German Users." Ph.D. dissertation, Dublin City University.

Shreve, Gregory. 2000. "Translation at the Millennium: Prospects for the Evolution of a Profession." In *Paradigmenwechsel in der Translation. Festschrift für Albrecht Neubert zum 70. Geburtstag*, edited by Peter A. Schmitt, 217–234. Tübingen: Stauffenberg Verlag.

Singh, Nitish, and Arun Pereira. 2005. *The Culturally Customized Web Site*. Burlington, MA: Elsevier.

TAUS. 2013. *Translation Technology Landscape Report*. De Rijp, the Netherlands: TAUS.

Thicke, Lori. 2011. "Improving MT results." *Multilingual*, January/February, 37–41.

Wu, Yonghui, Mike Schuster, Zhifeng Chen, Quoc V. Le, Mohammad Norouzi Wolfgang Macherey, Maxim Krikun, Yuan Cao, Qin Gao, Klaus Macherey, Jeff Klingner, Apurva Shah, Melvin Johnson, Xiaobing Liu, Łukasz Kaiser, Stephan Gouws, Yoshikiyo Kato, Taku Kudo, Hideto Kazawa, Keith Stevens, George Kurian, Nishant Patil, Wei Wang, Cliff Young, Jason Smith, Jason Riesa, Alex Rudnick, Oriol Vinyals, Greg Corrado, Macduff Hughes, and Jeffrey Dean. 2016. "Google's Neural Machine Translation System: Bridging the Gap between Human and Machine Translation." https://arxiv.org/abs/1609.08144

GLOSSARY

ATA American Translators Association

CAT Computer-Assisted Translation refers to a human translator's concurrent use of software to assist in and accelerate the translation process. As an aid to a human, it is distinct from MT.

Controlled language Language limited to relatively few specified words and grammatical structures, such that they can easily and quickly be learned for a particular professional domain, such as jet engine maintenance and repair. Most common are varieties—usually devised by international corporations—of controlled English and controlled French.

Explicitation A translator's addition of information that readers of the source text would infer but that readers of the target text need explained or made explicit. Explicitation may be accomplished with the addition of a word, phrase, sentence, or footnote.

FIT Fédération Internationale des Traducteurs (International Federation of Translators)

Globalize To prepare a document for use around the world, without translation or localization. Documents rendered in a controlled language are typically intended for globalized or internationalized use. For example, an instruction manual for airplane maintenance rendered in controlled Global English would be intended to be used by airplane mechanics anywhere in the world. See also "controlled language," "Global English," and "internationalize."

HT Human translation refers to any parts of the translation process done by a human rather than a machine. See also "MT."

Internationalize To undergo the process whereby one renders a product's features culturally neutral so that adaptations to various locales and languages can follow.

L1 A speaker's or writer's first language or mother tongue.

L2 A speaker's or writer's second language (not learned as a native). Third and fourth learned languages are numbered L3 and L4 respectively.

Localize To prepare a document or product for the culture and language of a particular *target* locale (e.g., Flanders, Quebec, southern Cyprus). Localization includes *translation* of language but is broader and often typically includes attention paid to adapting to cultural expectations for images, colors, level of formality, etc.

LSP Language service provider, such as a translation company or technical documentation firm

MT Machine translation is the substitution of a computer translation instead of human translators for the early stages of the translation process. It is distinct from CAT. For accurate interpretation, it is coupled with human post-editing.

PEMT Post-editing machine translation. See "post-editing" and "MT."

Post-editing Having a human translator complete the translation process begun by less-refined machine translation. However, increasingly sophisticated software is being developed to undertake the editing stage of the process as well. See also "MT."

Source language The original language of a text to be translated and localized. See also "target language."

Source text The text in the original language, before translation and localization. See also "target text."

Target language The language to which a text is translated and localized. See also "source text."

Target text The text in the translated and localized language. See also "source text."

Terminology management The selection and consistent use of terms to avoid confusion. For example, using synonyms (e.g., arrow, pointer) for the same item or part in a set of instructions can mislead a reader to think that two different items are involved. Terminology management dictates choosing just one and using it throughout the document.

TM Translation memory draws from computer databases storing previously approved translation equivalencies (typically phrases), in order to speed up translation by not having to translate all phrases in a document afresh.

Translate To render meaning from a *source* language to a *target* language. Effective translation finds suitable equivalencies of expression, rather than mere substitutions of words, as word-for-word translation typical results in incomprehensible or confusing and unnatural renderings. Translation between languages is not to be confused with the meaning of translation as sometimes used by those describing risk communication (e.g., A. J. Maule in "Translating Risk Management Knowledge," Liza Potts in *Social Media* *in Disaster Response*). In such usage, a message is relayed for different audiences but typically in various registers of the same language.

Translation brief A description of a text's audience, purpose, and attending situation, along with instructions to the translator. A brief allows translators to understand a text's contexts, thereby allowing translators to render a text's meaning more accurately, without resorting to guesswork.

INDEX